A geographical information system and satellite remote sensing to plan for aquaculture development:

A FAO - UNEP/GRID

udy in Costa Rica

ervice

FAO FISHERIES TECHNICAL PAPER 287

FOOD AND AGRICULTURE ORGANIZATION OF THE UNITED NATIONS
Rome, 1987

The designations employed and the presentation of material in this publication do not imply the expression of any opinion whatsoever on the part of the Food and Agriculture Organization of the United Nations concerning the legal status of any country, territory, city or area or of its authorities, or concerning the delimitation of its frontiers or boundaries.

ISBN 92-5-102575-4

PREPARATION OF THIS DOCUMENT

This document was prepared as a pilot study to call attention to applications of Geographical Information Systems and satellite remote sensing to plan for aquaculture development. Comments are welcome. Information about related on-going work and publications can be obtained by writing to the Chief, Inland Water Resources and Aquaculture Service (FIRI), FAO, Via delle Terme di Caracalla, 00100 Rome, Italy.

The study was a cooperative activity between the FAO Inland Water Resources and Aquaculture Service and the Global Resources Information Database project of UNEP. The Government of Costa Rica contributed personnel time for the ground verification, advice, and background information.

Distribution

FAO Fisheries Department
FAO Regional Fishery and Aquaculture Officers
Marine Sciences (General)
CIFA
IPFC
COPESCAL
UNEP/GRID
ADCP Regional Aquaculture Networks
Coastal and inland waters fishery and aquaculture projects
Inland Fisheries

For bibliographic purposes, this document should be cited as follows:

Kapetsky, J.M., L. McGregor and H. Nanne E., 1987 A geographical information system and satellite remote sensing to plan for aquaculture development: a FAO - UNEP/GRID cooperative study in Costa Rica. FAO Fish.Tech.Pap., (287):51 p.

ABSTRACT

Planning for aquaculture development has been hindered by a lack of comprehensive information on the land, water, human and economic resources available at national, sub-national, and ecosystem levels of organization. The objective of this study was to assess the capabilities of a Geographical Information System (GIS) and satellite remote sensing to rapidly provide synoptic information to plan for aquaculture development.

The study area was the Gulf of Nicoya on the Pacific coast of Costa Rica. Three kinds of aquaculture development opportunities were evaluated in terms of optimum locations and land and water surface areas available: (1) culture of molluscs in intertidal and subtidal areas as well as suspended culture of molluscs and cage culture of fishes, (2) extensive culture of shrimp and fish in existing solar salt ponds, and (3) semi-intensive shrimp farming along the gulf shoreline outside of mangroves. Criteria for evaluation included salinity, infrastructure, and water quality for all culture types; bathymetry, shelter from winds and waves, and security for intertidal, subtidal and suspended cultures; locations of existing salt ponds and availability of shrimp-post larvae for extensive shrimp farming; soil suitability, distance from fresh and saltwater, and land uses for semi-intensive shrimp culture.

The results are discussed in terms of accuracy, supplemental information to expand and refine the GIS, and ground and water verification. Geographical Information Systems and satellite remote sensing are assessed as to their advantages and constraints as tools to plan for aquaculture in developing countries.

An annex deals with satellite remote sensing of water quality in the gulf.

CONTENTS

1. INTRODUCTION

1.1 Background

One of the fundamental problems in planning for the expansion of aquaculture is to accurately assess the land, water, economic and human resources available for development. Comprehensive planning for aquaculture development is still uncommon. One of the reasons that such planning is not routine is that the means to synoptically integrate and analyze diverse kinds of resource, environmental and economic data have been lacking. Therefore, there is a need for an approach which can be used to rapidly identify areas of a country, or of smaller administrative or ecological units, suitable for various kinds of aquaculture as an aid to development planning. The means to facilitate such evaluations are available: specialized computer software which enables storage, manipulation and analysis of large quantities of data of diverse kinds having a common locational or geographic base. With such automation reporting can be in the form of tabular data, graphics, and most importantly, geographically coordinated maps. These are called Geographical Information Systems (GIS). Basic principles of Geographical Information Systems for land resources assessment have been explained by Burrough (1986), and many natural resource and environmental applications have been found for this technology. These have ranged from county-level decision making (American Farmland Trust, 1985) to country-wide inventories for developing countries (Conant, et al., 1983).

1.2 Objectives

Our object was to assess a Geographical Information System, supported by satellite remote sensing, as a tool to plan for aquaculture expansion in developing countries. By employing GIS technology we wanted to answer such basic questions for planners as:

- What opportunities exist for various aquaculture developments as defined by kinds and quantities of land, water and infrastructure?

- Among the opportunities, where are most promising locations for aquaculture development and what amounts of land and water are available?

By employing GIS and remote sensing technologies we sought advantages not only in time and costs, but also in a more comprehensive and integrated treatment of aquaculture development criteria than is usually possible by employing manual analytical and map-making techniques.

The Gulf of Nicoya was selected as the study site for a combination of reasons. Firstly, worldwide, aquaculture is developing rapidly in estuarine and coastal areas where there are many other competiting uses of land and water. Therefore, knowledge of the locations and areal expanses most apt for aquaculture are essential, if development is to proceed in a timely manner. Costa Rica, and the Gulf of Nicoya within it, are representative of such a situation. Secondly, a prior feasability study on mollusc culture and the extensive and semi-intensive shrimp farming already underway in the gulf area provide an opportunity to check some of the results of the present study against existing information.

1.3 GIS and Satellite Remote Sensing for Aquaculture Development

There are few examples of GIS applications in aquaculture in the literature. Mooneyhan (1985) showed how a number of physical factors - distance to water of a specified surface area, to a settlement and a roadway as well as location within a given vegetation type - could be merged to define optimum sites for aquaculture. This, however, was a simulation devised as a training exercise, and therefore required field testing. The present exercise is the conceptual decendant of that simulation.

Satellite remote sensing applications to aquaculture are more plentiful than those of GIS, but not numerous. For example, Travaglia and Lorenzini (1985) have used the Landsat Multispectral Scanner (U.S.A satellite sensor of 79 m resolution) to inventory commercially harvested algae and monitor their growth in a coastal lagoon. Loubersac (1985) used simulated SPOT (French commercial satellite sensor) data to demonstrate the capabilities of a high-resolution (10 or 20 m) satellite sensor for aquaculture siting.

Kapetsky (in press) illustrated the use of Landsat Thematic Mapper high resolution (30 m) data for inventory and classification of small water bodies for aquaculture development and fisheries management.

Cheney and Rabanal (1984) discuss aquaculture siting, design, engineering and management in relation to remote sensing, but in a context much broader than the use of satellite-borne sensors.

1.4 The Gulf of Nicoya

The Gulf of Nicoya is on the Pacific coast of Costa Rica at latitude 10 N. Its length is about 80 km. The gulf opens broadly onto the Pacific in its lower stretch, but the upper area is relatively narrow and enclosed (Fig. 1).

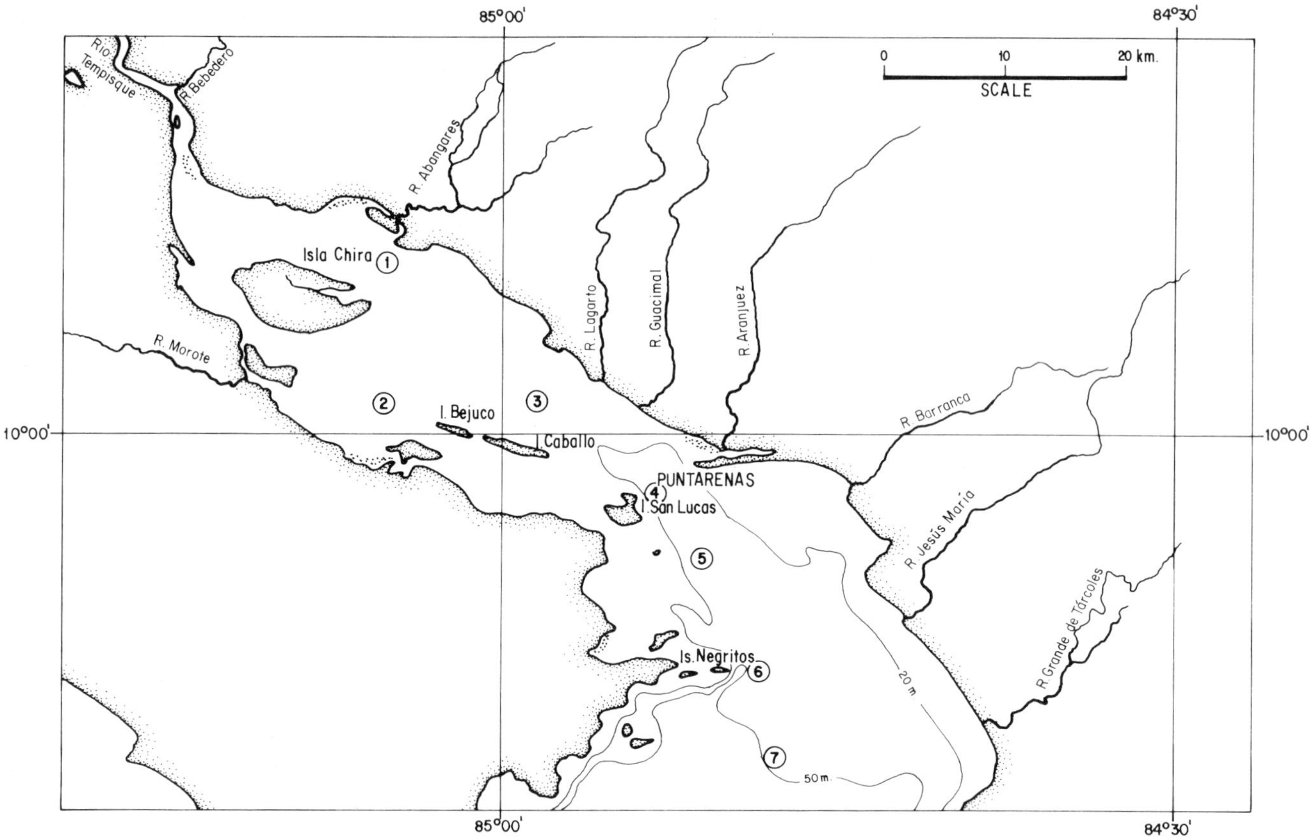

Figure 1. General features of the Gulf of Nicoya. Circled numbers are stations where Peterson (1960) collected salinity and other data.

The surrounding area is mainly farm and pasture land, and the only large population centre is Puntarenas; however, one of the large inflowing rivers, the Tarcoles drains an area in which about 70 percent of the nation's population resides (Rodriguez et al., 1978).

Much of the shoreline of the upper gulf is mangrove-fringed alternating with higher relief.

The warmest months are March, April and May (average about 29°C) and the coolest months are November, December and January (averages 26-27°C). Rainfall, totaling about 2 100 mm/y, is seasonal and occurs mainly from May through November. Strong N and NE winds are associated with the dry season (Peterson, 1960).

1.4.1 The aquatic environment

The outer portion of the gulf is relatively deep, up to 200 m, decreasing northward to about 20 m except in the vicinity of Puntarenas where there is a depression of about 40 m (Fig. 1). In the inner gulf depths are mainly less than 10 m. There are extensive intertidal areas mainly along the eastern shoreline and broad shallow areas mainly NW and S of Chira Island.

The tides in the gulf are semi-diurnal with a spring tide range of 2.8 m and a neap tide range of 1.9 m (Great Britain, Hydrographic Department, 1984).

Salinities are influenced by the seasonality of rainfall and runoff. Surface waters of the inner gulf, freshened by rainfall and runoff, flow southward on the eastern side of the lower gulf. This outward flow is compensated by an inward flow of more saline water on the western side of the gulf and on the bottom on the eastern side (Voorhis et al., 1983). Rainy season salinities in the upper inner gulf have been as low as 15 ppt. In the dry season surface salinities are equal to or near that of seawater (Peterson 1960).

Water temperatures are influenced both by winds and rainfall. Average monthly surface temperatures have varied from 30.3 in May to 25.8 in October. Temperature, like salinity and oxygen, is stratified in the rainy season in the upper gulf, but breaks down in the dry season with winds as a contributing factor (Peterson, 1960).

The Gulf has 8 major drainage basins. The most important rivers in terms of volume of flow are the Tarcoles and Barranca in the outer gulf, and the Tempisque, Morote, Abangares and Lagartos Rivers of the inner gulf (Fig. 1). Discharge rates vary widely from year to year. From 1981/82 to 1984/85 the annual volume of flow of the Barranca and Tarcoles Rivers varied by factor of about 1.4, the Abangares by 1.8, the Lagartos and Morote by 2.4 and the Tempisque by more than 3. The Nicoya peninsula is relatively dry and therefore, most of the freshwater inflow to the gulf is on the N and E sides.

1.4.2 Water Quality

Differences in dissolved oxygen in the upper zone suggest substantial primary production during the dry season and inputs of organic material from surrounding mangrove forests during the wet season. The lowest levels of surface and bottom dissolved oxygen were 67 percent and 60 percent of saturation in the upper gulf zone, and somewhat higher in the middle zone (Epifanio, Maurer and Dittel, 1983).

Relatively high levels of ammonium the year around in the mouth of the Tempisque River have been associated with agricultural activities in the basin, and lowered oxygen and high ammonium in the vicinity of Puntarenas are probably due to untreated sewage from the city which has a population of about 50 000 (Epifanio, Maurer and Dittel, 1983). Klemas et al. (1983) mention that the Tempisque River has an extensive agricultural watershed and could carry pesticides.

Trace metal concentrations in sediments and invertebrates have been studied by Dean et al. (1986) with the conclusion that the trace metals in the Gulf and its organisms are more likely attributable to the region's geology than to anthropogenic sources.

Epifanio, Maurer and Dittel (1983) suggest that nitrogen-laden Pacific waters are important in supporting production in the upper gulf especially in the dry season. Permanent stratification in the lower gulf prevents vertical mixing there, but the mid and upper gulf zones are fertilized by a kind of permanent upwelling of Pacific waters driven by the positive estuarine character of the gulf.

Klemas et al. (1983) have used Landsat MSS data combined with oceanographic observations to study water dynamics near Isla Negritos and the mouth of the Rio Grande de Tarcoles. These results as well as other aspects of water quality related to remote sensing are discussed more thoroughly by Jaquet (Annex 1).

1.4.3 Flora and fauna

The diatoms and the dinoflagellates dominate the phytoplankton of the gulf (Hargraves and Viquez, 1985). Red tides have been observed in the Gulf de Nicoya for some time in different seasons, but have been studied only by Hargraves and Viquez (1981). Cochlodiium catenatum was the principal organism, but it apparently does not produce toxic metabolites. All of the red tides have been in the mid or lower gulf.

The diversity and distribution of soft-bottom benthos have been investigated by Maurer and Vargas (1984). Polychaetes dominated in numbers and by numbers of species.

Because of the importance of the fishery of the gulf there have been several studies of fishery resources including crabs (Dittel, Epifanio and Bautista, 1985) and finfish abundance and distribution (Bartels et al., 1983; Bartels et al., 1984).

1.4.4 Aquaculture

Shrimp culture

The first efforts, in 1974, to develop shrimp culture in Costa Rica were on the N shore of the gulf near Chomes. By 1980 there were 112 ha of ponds, most of which were constructed in mangroves. The exposure and subsequent oxidation of the mangrove soils caused acidic conditions and resulted in low production. At times there was an insufficient supply of post-larvae for culture. Eventually the company went bankrupt. Recently, a new company has taken over the facility and has 45 ha of ponds in production. Seedstock is supplied naturally, but there are plans for a hatchery. Because of past problems with siting shrimp ponds in mangrove areas and because of an awareness of the need to conserve mangrove ecosystems for fisheries, new construction for the development of shrimp farming is to be outside of mangrove areas.

On the shores of the gulf there are about 120 operators of ponds used for solar salt production. Because of overproduction of salt, and because salt production is difficult in the rainy season, there is interest in using the ponds for extensive polyculture of fish and shrimp (Phillips, 1985). Towards this end there is a study to evaluate the abundance of young shrimp. External assistance on culture methods is being sought from neighboring countries. On the N shore of the gulf there are now 5 producers of shrimp in solar salt ponds.

Mollusc culture

As part of a larger study on the potential for mollusc culture for the entire country, Glude (1981) surveyed parts of the gulf. He concluded that there are possibilities for oyster, mussel and clam culture. Among the advantages of the gulf area are that it is nutrient-rich, but believed to be little affected by pollution. Oyster farming trials with Crassostrea gigas have been underway with promising results and there has been limited work on C. rhizophorae. Two species of mussels are fished in the gulf as are two species of ark shell clam, but applied research is required to test the technical feasibility before commercial culture activities can be begun.

2. MATERIALS AND METHODS

The kinds of aquaculture which appear to have the most promise for development in the Gulf of Nicoya area are (1) mollusc culture in intertidal and subtidal areas and in deeper waters by suspension from rafts or longlines, the latter also corresponding to areas for the culture of fish in cages, (2) extensive shrimp farming in existing solar salt ponds, and (3) semi-intensive shrimp culture outside of mangrove areas.

We focused our study on the inner area of the gulf, where it was obvious from the outset that more aquaculture opportunities were to be found than in the deeper, ocean-like outer gulf. We define the inner gulf (polygon GULF in our GIS) as the area enclosed by a line northward from the eastern end of Isla Negritos Afuera to the Puntarenas peninsula, and a line bounding the south shorelines of both Islas Negritos westward to the shoreline (Fig. 1). All of our results reported as surface area are for the inner gulf.

Our general approach was to (1) specify the kinds of information required to plan for the three kinds of aquaculture development mentioned above, (2) extract and compile the tabular data, maps, and charts containing the information for aquaculture planning, (3) compile the GIS by digitizing the map and chart data and by direct entry of tabular data, (4) manipulate and analyze the data by merging the relevant data sets, and (5) report the results in map and tabular form.

2.1 Computer Software and Hardware

The software used to compile, analyze and report the data was Earth Resources Applications Software (ELAS) (Whitley et al., 1981; Graham, et al, 1985). This software processes geographically referenced data including remotely sensed data from a variety of satellite and airborne scanners such as the Landsat Thematic Mapper data used in this study. Digitized data from maps and tables also can be processed and merged with the scanner data.

The software has been adapted to run on a variety of commonly available computers. A simplified version runs on personal computers. ELAS is under continuous development by the Earth Resources Laboratory of the US National Aeronautics and Space Administration and by the Global Resource Information Database of the UN Environmental Programme.

The hardware used for the analyses included an Perkin-Elmer 3241 computer, and maps and charts were digitized on a CALCOMP 6000. The analyses were made at the Global Resources Information Database (GRID) Processor Facility in Geneva, Switzerland. Disc and tape units, terminals, monitors, and printers were those common to any large image processing or Geographical Information Systems facility.

2.2 Criteria to Assess Aquaculture Potential

The criteria used to assess aquaculture development opportunities are listed in Table 1. Some criteria, such as salinity and infrastructure are important for all types of coastal aquaculture development. Others, for example, site acquisition costs, pertain only to semi-intensive shrimp farming in this instance.

Salinity

Salinity is an important physiological criterion for all cultured animals. Too high or too low salinity affects growth, and if extreme can be lethal. Low salinity (< 10 ppt) can produce an off-flavour in shrimp (Clifford, 1985). We portray the spatial and temporal distribution of salinity with the objective of locating the areas with optimum salinity regimes for molluscs and shrimps.

Minimum annual surface salinities at 9 stations (Fig. 1) were established from data tabulated by Peterson (1960) based on observations made from April 1952 to December 1957 (Table 2.)

Minimum surface salinities were interpolated between stations over most of the gulf are and extrapolated from Station 1 towards the mouth of the Tempisque River using the distance (DIST) function in ELAS.

Bathymetry

Bathymetry, with a data base datum at mean lower water springs, was obtained from the Gulf of Nicoya nautical chart at 1:100 000 scale (Instituto Geográfico Nacional, Servicio Geodesico Interamericano, 1984) and digitized in three depth categories to reflect the general requirements of the various kinds of mollusc culture methods (Quayle, 1980; Angell, 1986) and for cage culture of fish. The distinction between intertidal and subtidal has important implications for fouling of culture installations and accessibility of cultured organisms to predation.

Bathymetry		Culture Method
Intertidal	>0 m	[Bottom, or off-bottom culture
		[on stakes or racks using
Subtidal	0 - 5 m	[trays, sticks or strings
Suspended	>5 m	[Rafts, long-lines or cages

Table 1

Criteria used to identify opportunities for mollusc, fish and shrimp culture in the inner gulf area. X's indicate the individual criteria which were applied to each type of culture activity

Main Criteria and Sub-Criteria	Organisms and Culture Methods: Molluscs		Fishes	Shrimps	
	Intertidal	Subtidal	Suspended	Extensive	Semi-intensive
Salinity	X	X	X	X	X
Infrastructure	X	X	X	X	X
- Villages					
- Roads					
- Ferries					
- Processor					
- Electrical service					
Land Uses					
- Water Quality	X	X	X	X	X
- Site acquisition costs					X
- Site development costs					X
- Mangroves	X	X			X
Bathymetry	X	X	X		
Shelter and Security					
Shrimp post-larvel density				X	X
Proximities					
- Perennial rivers					X
- Saltwater					X
Soils					X

Table 2

Summary of surface salinity data from 9 stations acquired by Peterson (1960)

Station	Minimum salinity (ppt)	Observations per station	No. of calendar months sampled
1	15	42	12
2	20	14	11
3	24	14	10
4	23	45	12
5	25	42	12
6	29	12	11
7	26	37	11
8	26	37	11
9	29	7	7

Shelter and Security

The winter wind regime of the gulf is dominated by Papagayo northerlies. These winds blow from one-half to four or five days at a time, and often reach gale velocities Peterson (1960). Analysis of data from 1970 shows that the most frequent winds at Puntarenas are from the S (38 percent), but the N winds have the highest average velocities. Examination of raw wind data from Puntarenas from June, 1978 to May, 1979 (obtained directly from the Instituto Meteorologico Nacional, San José) showed maximum average hourly wind velocities of 37 kph. Using a relationship between wind velocity, fetch and water depth (U.S. Army Coastal Engineering Research Center, 1977), the theoretical wave height of 37 kph winds in water 3 m deep with a fetch of 2 nm (3.7 km) would be 0.3 m and in water of 6 m depth, 0.4 m. These wave heights were deemed to be within the limits for intertidal, subtidal and suspended cultures.

Shelter from wave damage and wind-induced turbidity was established by digitizing a 2 nm (3.7 km) border N and S of each land mass on the same chart as used for bathymetry.

Implicit in shelter as siting criterion also is the requirement that culture sites have to be monitored to detect breakage of structures and to prevent theft. Two nautical miles (3.7 km) is the greatest practical distance from shore for such surveillance and for timely interventions against damage and poachers.

Land Uses

Land uses in broad categories - water, coniferous forest, deciduous forest, range, pasture, marsh, cropland, shrimp ponds, salt works ponds, and mangroves as *Avicennia* and *Rhizophora* - were interpreted from the 15 April 1985 Landsat Thematic Mapper (Row 053 Path 016) data and verified by a ground survey in July, 1986.

Information to assess water quality constraints on aquaculture development is scant. Therefore, we used the proportions of land uses in individual river drainages to infer the relative quality of water of the rivers emptying into the inner gulf. For this purpose the gulf basin river drainages were defined and digitized. Then, river drainage boundaries were merged with the land uses interpreted from the Landsat TM image. The inner gulf drainages, the Morote, Bebedero and Abangares (the latter also including the Lagarto and Guacimal Rivers), were entirely or mostly included in the satellite image, but only the lower part of the Tempisque was covered (Fig. 2). In addition to the river drainages, the small islands of the gulf were grouped together to form a single land-use unit, and Chira Island made up another. Most of the aquaculture development in the gulf area is likely to take place in the drainage basins included on the satellite image.

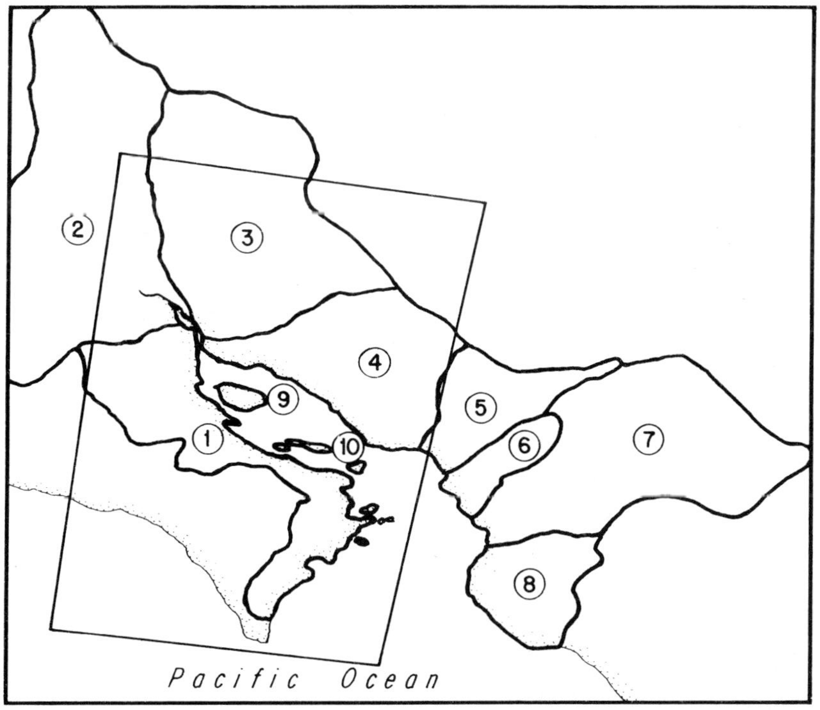

Figure 2. The river and island drainages of the Gulf of Nicoya. The rectangle shows the area coverage of the 15 April 1985 Landsat Thematic Mapper image. The river drainages are: (1) Morote (2) Tempisque (3) Bebedero (4) Abangares (5) Barranca (6) Jesus Maria (7) Tarcoles (8) Herradura. The island drainages are: (9) Chira (10) small islands grouped together.

Interpretation of water quality in each river basin in relation to land use was: (Best) Mangrove > Forest > Range > Pasture > Cropland (Worst).

For semi-intensive shrimp farming, land uses were interpreted in three ways: as suggesting local water quality (e.g., a high proportion of agriculture in a given area signifies the possibility of pesticide pollution), in terms of site acquisition costs (e.g., land developed for crops would cost more to buy than rangeland), and as indicative of site development costs (e.g., forested land would be more expensive to prepare for ponds than land already cleared for pasture).

Identification and quantification of mangroves were important for several aspects of the study. Mangrove distribution was used to assess development prospects for semi-intensive shrimp farming. Because of past difficulties with acid soils at the shrimp farm at Chomes on the N side of the gulf and because of the desire of the government to conserve mangroves for their fishery benefits, sites for semi-intensive shrimp farms were considered only if outside of mangroves.

Mangrove detection and quantification also were important to assess densities of post-larval shrimp for extensive culture of shrimp in solar salt ponds and to assess opportunities for the development of mangrove-compatible aquaculture.

Infrastructure

The objective was to locate and quantify the infrastructure important for aquaculture development in the gulf area. This includes:

- The road and water transportation system in the immediate vicinity of the gulf.

- Villages and district centres as indicative of sources of local labour and supplies.

- Electrical service to power aquaculture operations.

Roads and/or water routes are required to transport cultured products to processors and markets and by which to receive supplies necessary for culture operations. Especially important from a cost point of view is distance and the travel time. In this regard the gulf transportation network is made complex by the need to use ferries between the N and S shores.

Proximities are important in two ways: site development costs and culture operations costs. Close proximity to a road or landing place will mean that site development costs for road building will be less than for a site more distant. Likewise, close proximity to a facility for processing and marketing aquacultured products, to a large town for supplies, or to a feed mill will mean that transport costs will be less than from a more distant site. For example, in the semi-intensive culture of shrimp the largest single operating cost can be feed. A 10 ha pond could require about 25 tons of feed in a 110 day growing cycle. Thus, feed haulage costs as reflected by the distance between the feed mill and the shrimp farm could be an important factor in site selection.

An additional infrastructure particularly important for locating semi-intensive shrimp farming is the availability of electricity from a central grid. Generally, electricity is a cheaper source of energy for pumping water than diesel motors when capital costs and maintainance are taken into account. Thus, a location with electricity nearby could represent lower operating costs than one dependent on diesel or diesel-electric pumping for water circulation. Likewise, if cultured products are to be held at farm sites, then electricity from the national grid to supply cold stores or for ice-making is an advantage. If electrical service is available, but unreliable, then it is much less important as an aquaculture development criterion because generators have to be purchased and maintained to provide electricity when the normal service is interrupted.

Puntarenas, of about 50 000 inhabitants, provincial capital, fishing port, fish processing and population centre, was used as the focal point for road and water transport. In order to reach other large inland population centres it would be necessary to pass close by Puntarenas. Thus, proximity from potential aquaculture sites to Puntarenas is indicative of distances to reach the infrastructure necessary to support aquaculture development both in the province and in the country.

The road and water transportation network, locations of villages ("caserios") and district centers ("cabeceras de distritos") were obtained from the physical-political map of Costa Rica at 1:500 000 (Instituto Geográfico de Costa Rica, 1983). In some cases the road network was updated with the Landsat TM imagery. This information was digitized and an ELAS programme was used to measure the road distances from a number of villages and district centers to Puntarenas. ELAS also was used to generate distances over water to all points on the inner gulf from Puntarenas.

Soils

Soil physical and chemical characteristics are important for semi-intensive shrimp farming in purpose-constructed ponds. The general objective was to locate soils which are relatively impermeable, sufficiently compactible to be shaped into dikes, and which do not have acid potential. Soils were digitized from a map of soil association sub-groups for Costa Rica at 1:200 000 scale (Perez, Alvarado and Ramirez, 1978). Interpretations on suitability for ponds were made by H. Brammer, FAO Soil Resources, Management and Conservation Service (pers. comm., 1986), and based on his characterizations, the soils bordering the gulf were ranked in five categories (Table 3).

Table 3

Soil units bordering the Gulf of Nicoya and their interpretation for the development of shrimp ponds

Unit ID.	Rank (relative)	Unit Names
M3	1	Fluvaquentic Hapludoll
I24	1	Aquic Ustropept
I21	2	Fluventic Ustropept
I22	2	Fluventic Ustic Dystropepts
I33	2	Ultic Dystropept
I19	3	Typic Ustropept
I23	3	Lithic Ustropept
V1	4	Vertisols (Grumosols)
E5	4	Lithic Usterthent
I32	4	Ustic Dystropept
E6	5	Typic Sulfaquent

Solar Salt Ponds

Solar salt ponds can be used for the extensive culture of fish and shrimp, if properly managed. The objective was to locate and inventory the ponds around the inner gulf periphery, then to identify those most suitable for aquaculture development using as criteria salinity, availability of post-larval shrimp, and infrastructure.

Solar salt ponds are shown on 1:50 000-scale maps available from the Instituto Geográphico Nacional in San José, but the maps are quite old with latest updates for most in the late 1960's and 1970's. Therefore, we decided to take advantage of the availability of a recent Landsat TM image of the gulf (15 April 1985) to provide an up-to-date inventory of solar salt ponds 1 ha and larger in area. The salt ponds were detected on the basis of their spectral characteristics and were separated from shrimp ponds based on knowledge of the locations of the latter.

Proximities to Salt Water and Perennial Rivers

Fundamental criteria to locate opportunities for semi-intensive shrimp farming are the availability of saltwater and freshwater. In semi-intensive shrimp culture sufficient water has to be available for pumped circulation to maintain water quality. An exchange of 5 to 15 percent per day is usual (Clifford, 1985). But also, by varying proportions of fresh and saltwater (as would be required throughout the year because of changes in gulf salinity due to rainy and dry seasons) salinities in a narrow range optimum for shrimp growth, generally 15 to 25 ppt, could be maintained. Given a pond depth of 1 m, a salt water source at 33 ppt and a 10 percent water exchange, 390 m^3/d of freshwater per hectare of pond surface would be required to maintain salinity at 20 ppt, the middle of the range optimum for growth. Ten times that amount would be required to reach 20 ppt at the initial filling.

There are only 4 rivers for which annual volume of flow data are available for a number of years (data from Instituto Costaricense de Electricidades) and which also are included in the 15 April satellite image: Morote, Tempisque, Lagarto, and Abangares (Fig. 2). Of these, we did not select the Tempisque. The Tempisque flow is so large that near-surface salinities are likely to be well below optimum during the rainy season in the vicinity of its mouth. However, it is possible that an underlying saltwater wedge could be tapped for higher salinity water as suggested by the maximum difference between salinities at surface and at 5 m at Station 1, 9 ppt, as recorded by Peterson (1960).

In place of the Tempisque we added two smaller rivers, the Guacimal and Aranjuez. Volume of flow data were available only for one month, March, but this is one of the months of lowest flows and therefore indicative of minimum amounts of water available during the course of the year. The March flow of the Guacimal, the least of the two, would be sufficient to provide freshwater to dilute seawater for several hundred hectares of shrimp ponds under the conditions set out in the paragraph above, if about one-half of the flow were used for this purpose.

The maximum economically practical distance for the transport of saltwater or freshwater to ponds was assumed to be 1 km. Further, saltwater was assumed to be available landward as far as the inner extension of mangroves. Using a distance sub-programme of ELAS, a 1 km band was generated landward from the back of the mangrove to portray the maximum distance that would be economic for the transport of saltwater. Likewise, a 1 km band was generated on either side of each river to represent the area in which freshwater could be transported by gravity or pumping. The intersection of these two areas defines locations where both salt and fresh water are no more distant than 1 km from the back of the mangroves.

Density of Shrimp Post-larvae

The objective was to predict the locations on the gulf shoreline where post-larval shrimp are most likely to be abundant. Shrimp post-larvae have to be available for semi-intensive culture because supplemental stocking is essential to high productivity. If hatchery-reared post-larvae are not available, then they have to be collected from the wild. Thus, there is an advantage in knowing where shrimp post-larvae are most abundant so that "seed" fisheries can be developed. In Costa Rica shrimp farms with surface areas larger than 200-300 ha are permitted to obtain shrimp post-larvae from the wild only during the first year of operation, after which seedstock must come from hatcheries (Committee of Inland Fisheries for Latin America, 1986).

Because the cost of post-larvae can amount to about 30 percent of total operating and harvesting costs in semi-intensive shrimp farming (Hollin, 1985), any means which increases availability or improves survival can increase profitability. In this regard, proximity to high densities of seedstock could be an important siting factor in two ways. The first is that losses due to transport could be minimized, and shrimp would be in better condition when stocked than if carried long distances, thereby increasing survival. The second is that with the farm in an area of high-density of post-larvae there would be more post-larvae to be entrained into the water supply canals during daily pumping for water exchange thereby lessening stocking needs.

The economic viability of extensive shrimp and fish farming in solar salt ponds will depend mainly on the quantities of shrimp harvested. The fish likely to be present in the ponds will be of lower value than shrimp, even if more numerous. If solar salt ponds are dependent on self-stocking of fish and shrimp carried in on tidal currents, then, it follows that solar salt ponds located where shrimp post-larvae densities are naturally high will have a production advantage.

Several factors influence the abundance of shrimp post-larvae including substrate type and current velocity, but such information is incomplete for the gulf. Therefore, for this exercise we have equated the complexity of the water-mangrove interface with shrimp post-larval density based on the preference of P. vannamei for low salinities (Edwards, 1978; Hughes, 1985) and its association with mangroves and waterways (Edwards, 1977). We focused on P. vannamei because of its numerous advantages for culture including its hardiness, wide envrionmental tolerance and high survival in ponds (Clifford, 1985). It is a logical species to consider for culture in solar salt ponds because post-larvae are abundant during the rainy season when salt-making is most difficult.

We reasoned that the greater the complexity of the mangrove-water interface, (i.e., the longer the interface per unit of area), the more the shrimp habitat available, and the greater the density of shrimp post-larvae per unit of area. The complexity estimates were generated by ELAS sub-programmes SLIN and SLID for each mangrove zone around the gulf shoreline.

This approach has been used to explain shrimp, crab and fish yields as a function of shoreline complexity (Dow, 1982).

Other Criteria

In many instances water temperture could be an important aquaculture development criterion; however, in the case of the inner gulf, water temperatures are favourable for aquaculture and do not vary greatly. Peterson (1960) notes that surface temperatures reached 32.5°C in shallow areas near Chira Island, and the lowest temperature he encountered at 10 m depth at Station 1 from 1952 to 1957 was 25.3°C. Therefore water temperature was not used as a locational criterion, but could be important when considering the possibility of introducing an exotic species for culture in the inner gulf (e.g., Glude, 1981).

Topography is important to locate areas suitable for semi-intensive shrimp farming in relation to elevations favourable for pumping for water management. Usually, elevations higher than 3 m above mean high water would be uneconomic for pumping. However, the 1:50 000 maps available from the Instituto Geográfico Nacional in San José are contoured only at 10 m intervals whereas contours at intervals less than one-half of that would be requried to assess large areas for the feasibility of pumping.

3. RESULTS

First, the analyses of criteria of common concern for the development of all aquacultures are presented: water quality, infrastructure, and salinity. Then, opportunities for the development of each kind of culture are assessed in terms of their locations and areal expanses. These are (1) the culture of filter feeders in intertidal, and subtidal areas and by suspension, the latter of which also corresponds to areas suitable for cage culture of fish; (2) opportunities for extensive shrimp and fish farming in solar salt ponds; and (3) opportunities for the semi-intensive culture of shrimp outside of mangrove areas.

3.1 Water Quality in the Islands and River Drainages of the Inner Gulf

In this interpretation we associate mangroves with highly productive waters. Rangeland could be a positive influence on water quality, if it is not overgrazed, likewise pasture, if it is well managed. We associate with crops such negative effects on water quality for aquaculture as lethal and sublethal concentrations of pesticides, turbidity through soil erosion, and fluctuating quantities of runoff because of the lack of absorbative capacity of cultivated crops compared with natural vegetation.

The relative amounts of land uses among the four river and two island drainages of the inner gulf are shown in Figure 3. The grouped small islands are mainly forest and range without cropland. It can be assumed that good quality water drains these islands. Chira Island, too, would appear to have good quality water for aquaculture because it has relatively high proportions of mangrove, forest, and range, a relatively small area of pasture, and no cropland. Among the river drainages, the Morote and Abangares are relatively well forested, and have large proportions of range and little pasture or cropland. Thus, water quality would be expected to be relatively good. In contrast, the Bebedero and Tempisque drainages have higher proportions of crops and pasture and less forest and range, and therefore water quality may be somewhat lower than in the Morote and Abangares drainages. However, the results from the Tempisque have been influenced by having only the lower portions of the drainage included in the assessment (Fig. 2).

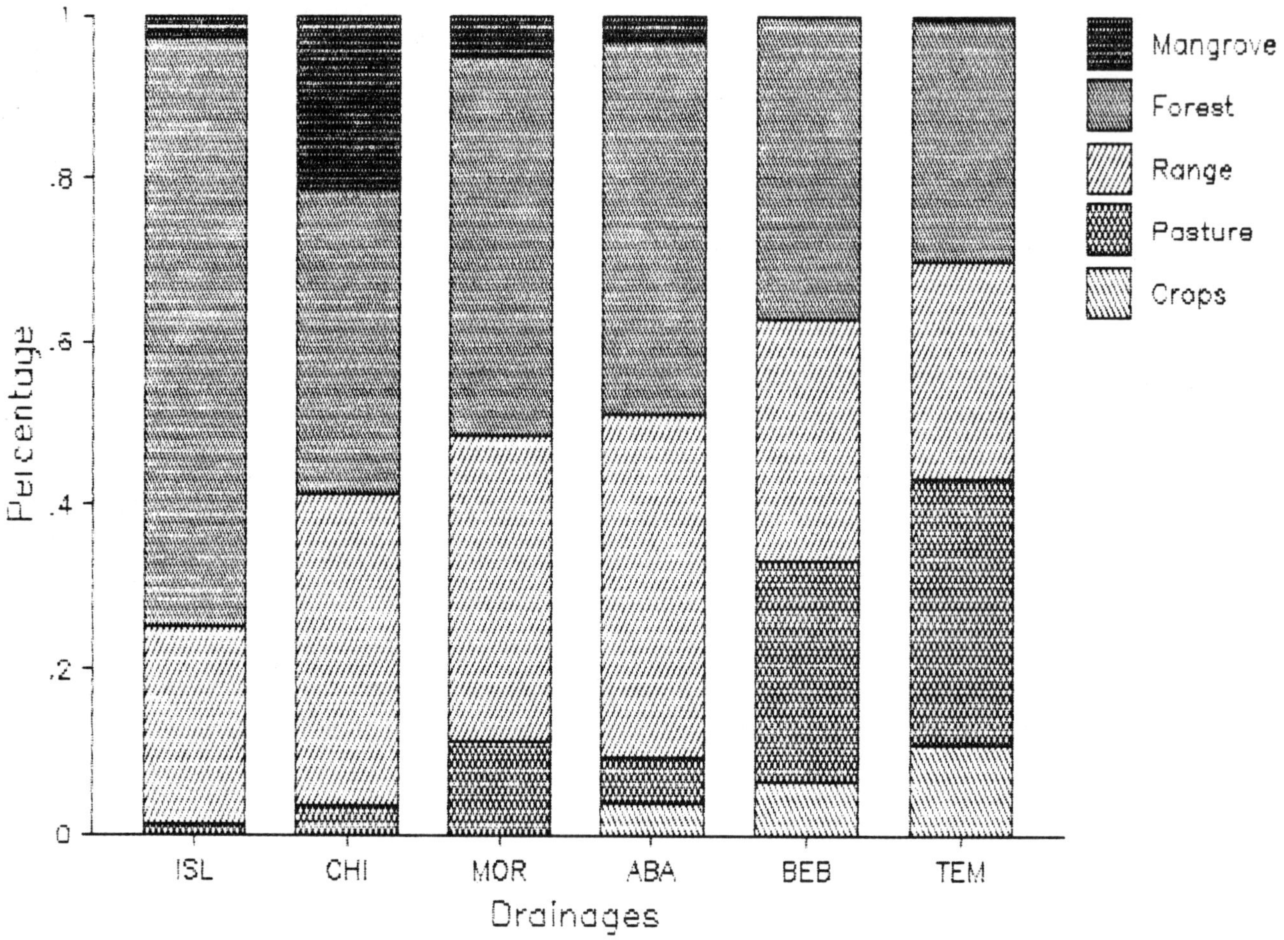

Figure 3. Land uses in the island and river drainages of the inner gulf. ISL - small islands grouped; CHI - Chira island; MOR - Morote; ABA - Abangares; BEB - Bebedero; TEM - Tempisque

Additional aspects of water quality interpretations using satellite data are presented by Jaquet in Annex 1.

3.2 Salinity

Interpretation of salinity information for aquaculture development purposes has to be general for several reasons. Firstly, salinities in the inner gulf area, from Puntarenas inward, are based on observations at only 4 stations (Fig. 1). Secondly, although there were numerous observations over the course of the annual cycle and among years at Stations 1 and 4 of the inner gulf (Table 2), the stations are in open waters, well away from the shoreline where the solar salt ponds are located and intakes for semi-intensive shrimp farming would be positioned. Nevertheless, a number of useful inferences can be made:

1. Minimum surface salinities at openwater stations increase from the upper inner gulf to the lower inner gulf as is shown by Figure 4 which is a interpolation of minimum values between stations. This is explained by the inflows of the Tempisque, Bebedero, Morote and other, smaller rivers. The distribution of

minimum salinities on Figure 4 is indicative of conditions which would be encountered by organisms used in cage, raft, or longline culture, if these were suspended near the surface. Culture structures set deeper would encounter higher salinities for more than one-half of the year as suggesteded by salinity data from Station 1 where salinity differences in the water column are most likely to be highest. The table below indicates the maximum amount by which salinities at 5 m and 10 m exceeded surface salinities at Station 1 on a monthly basis from 1952 to 1957 using data from Peterson (1960):

	Month											
	J	F	M	A	M	J	J	A	S	O	N	D
Depth	Salinity (ppt)											
5 m	0	1	0	0	0	7	9	1	8	2	2	5
10 m	0	1	0	0	0	14	12	3	12	3	6	6
n	1	3	2	5	3	3	4	1	5	2	3	6

2. Salinities inshore are likely to be lower during the rainy season than those portrayed in Figure 4 and in the the curves of Figure 5a to 5d because of the influence of rainfall runoff. Locally, close to the inflowing river mouths, salinities could be much lower. Based on average volume of flow data for recent years (data from Instituto Costaricense de Electricidades), the Abangares and Lagartos Rivers have about equal discharges, the Morote is twice as large as these, and the Tempisque flow is 6 times greater. Thus, the Tempisque and Morote outflows exert a greater influence on salinities locally than either the Abangares or Lagartos.

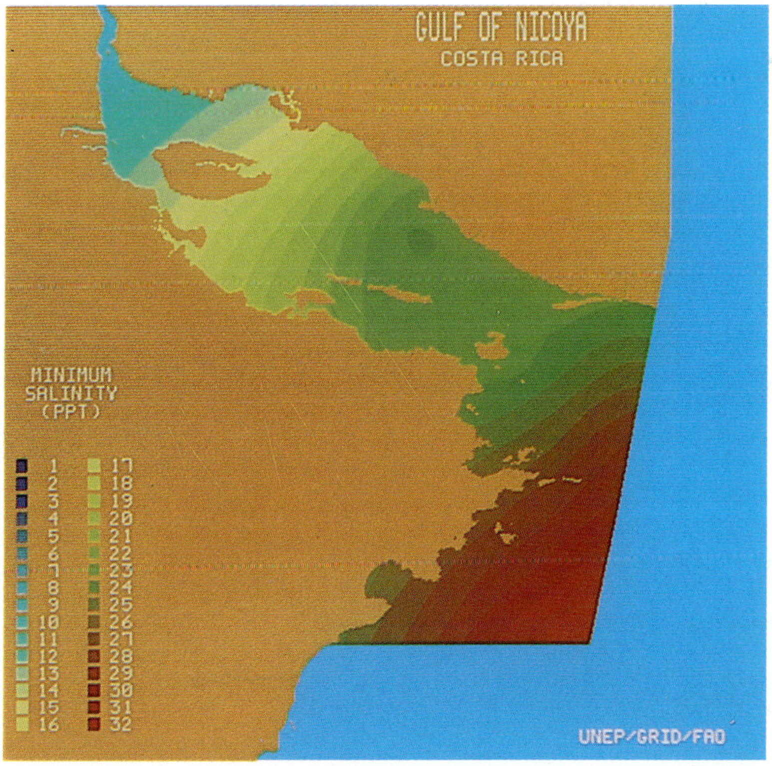

Figure 4. Minimum annual surface salinities interpolated between stations based on data from Peterson (1960). Station locations are shown in Figure 1. Salinities have been extrapolated from Station 1 to the mouth of the Tempisque River

3. Salinities during the rainy season are quite variable from year to year at the same stations, but vary much less during the dry season (Fig. 5a to 5d.).

4. Along most of the southern shoreline of the inner gulf salinities are likely to be higher than the range of 15 to 25 ppt optimum for shrimp grow-out for most of the year because of the lack of perennial rivers there, except for the Morote. By the same reasoning, the northern shoreline is expected to have salinities lower than those of the southern shore because of more perennial rivers (Fig. 1).

5. The upper inner gulf has the greatest annual range in surface salinity. The annual average salinity curves (Fig. 6) show this effect with the minimum salinity observed at Station 1 15 ppt while minimums at stations 2, 3 and 4 ranged from 20 to 24 ppt. Dry season salinities tend to converge at from 30 to 33 ppt among all 4 inner gulf stations.

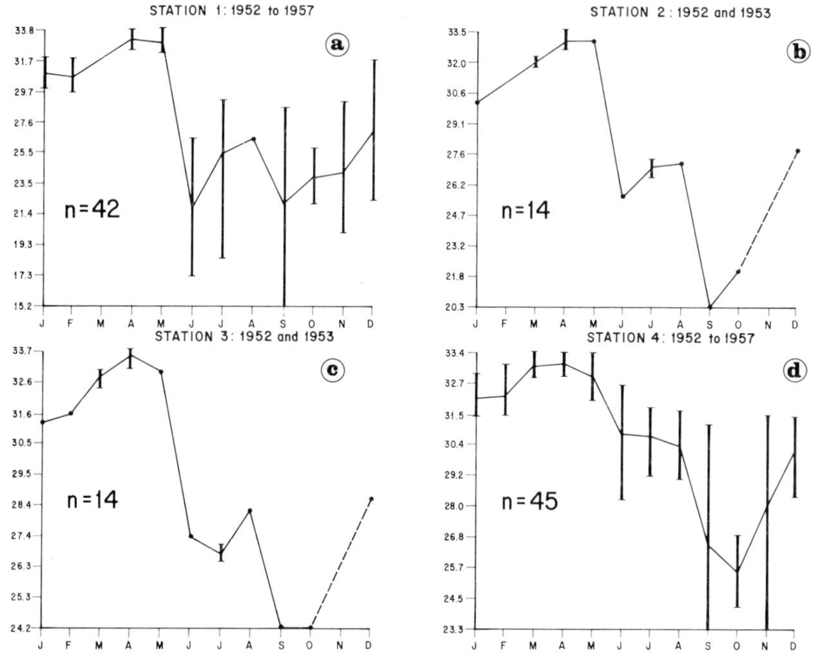

Figure 5. Mean monthly surface salinities and ranges at the 4 inner gulf stations. Note that Y-axes have different ranges. Station locations are on Figure 1.

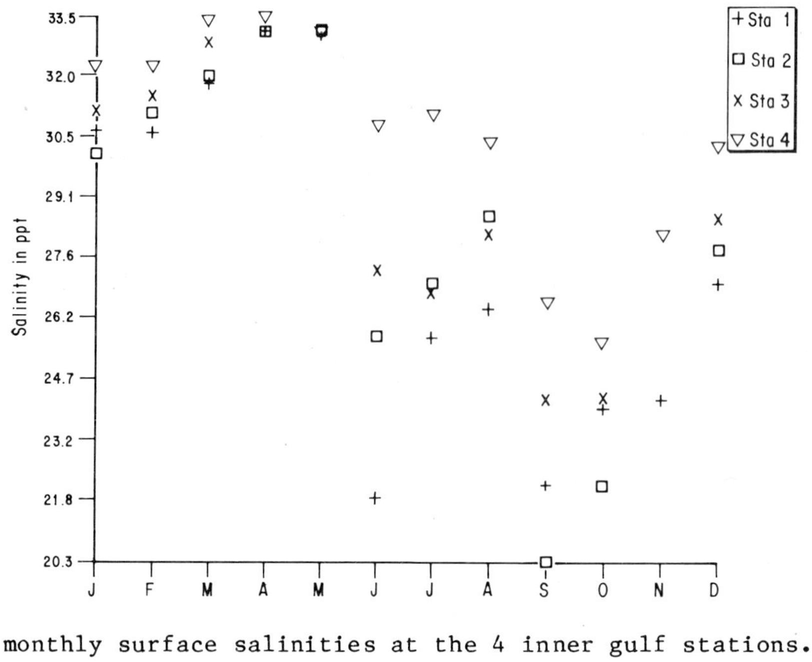

Figure 6. Mean monthly surface salinities at the 4 inner gulf stations.

Thus, in summary, optimum salinities for shrimp farming are most likely to be found on the north shore of the inner gulf near Chira Island (where the minimum salinity observed at Station 1 was 15 ppt), and in the vicinity of the mouths of the north shore rivers further outward.

3.3 Infrastructure

From the point of view of road transport, and with Puntarenas as the focal point, the south shore of the gulf is less accessible than the north shore even though distances are not longer (Table 4). This is because the south shore can be reached only by using ferries, one from Puntarenas to Cabo Blanco and the other across the Tempisque River at Puerto Moreno (Fig. 7).

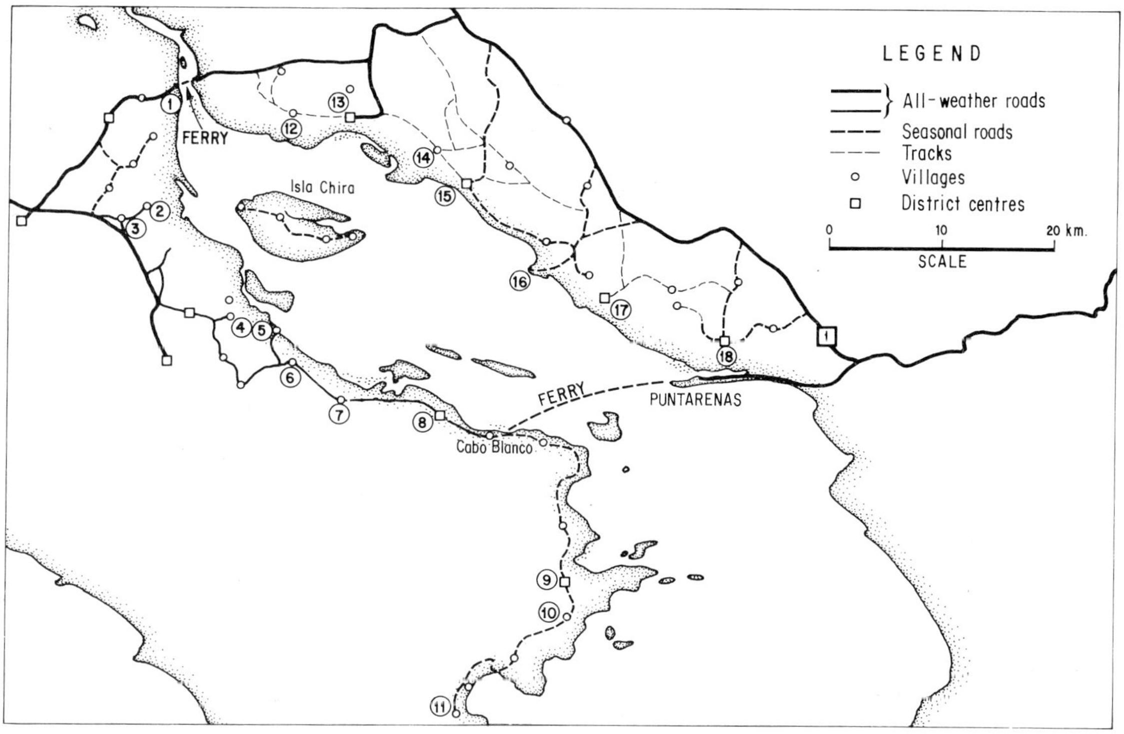

Figure 7. Infrastructure for aquaculture development in the inner gulf area. The villages and district centres indicate availability of labour and supplies. Distances to Puntarenas along the road-water network are shown in Table 4.

Table 4

Distances by road (including ferry routes) from district centres and villages around the periphery of the Gulf of Nicoya to Puntarenas. Numbers are keyed to Fig. 7 for the locations of the villages and district centres

No. on map	South Shore	km	No. on map	North Shore	km
1	Pto Moreno	115	12	S. Buenaventura	107
2	Pto Jesus	89	13	Colorado	103
3	Vigia	86	14	Abangaritos	85
4	Pto Thiel	70	15	Manzanillo	74
5	El Canjelito	61	16	Pta Morales	65
6	Corozal	55	17	Chomes	64
7	Jicaral	45	18	Pitahaya	30
8	Lepanto	31			
9	Paquera	49			
10	Curu	54			
11	Tambor	79			

Representative distances over water to Puntarenas are about 40 km from the westernmost part of the gulf at Puerto Moreno, 30 km from the eastern end of Chira Island, and 17 km from Puerto Morales (Fig. 7).

On the both the N and S shores there are a number of small villages and district centres evenly spaced along the road ways. Thus, local supplies and labour generally are readily available for aquaculture development all around the gulf periphery.

Electrification is an important siting criterion in consideration of capital and operating costs, if reliable, especially for pumping water for semi-intensive shrimp culture and for maintaining cold stores for all kinds of culture operations. Electrical service was available only along a small part of the S shore of the gulf, but more widespread on the N shore in 1979 (Fig. 8).

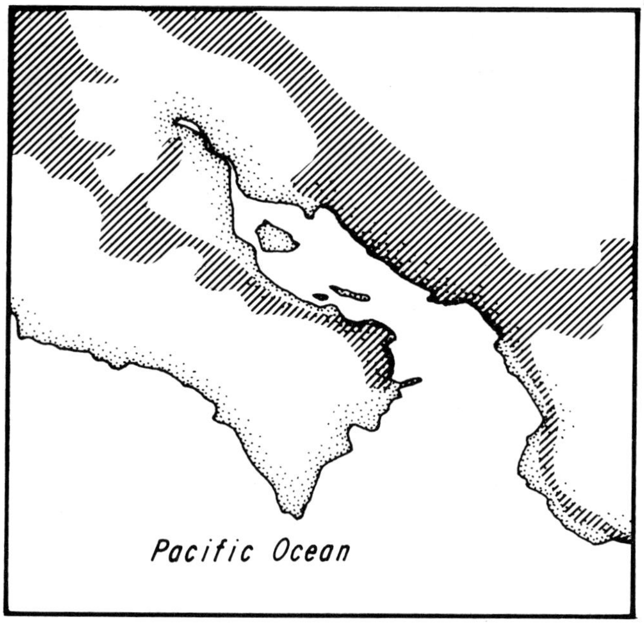

Figure 8. Areas with electrical service in 1979 (adapted from Fig. IV- 19 in Hartshorne et al., 1982). Because of the age of the data and the scale, the locations are indicative only.

3.4 Intertidal, Subtidal and Suspended Culture

The intertidal mudflats are defined as the area between the shoreline and the water interface at Mean Low Water Springs. In this habitat oysters would be grown directly on the bottom, or artificial substrate could be added. In the inner gulf, this area is relatively small, 3 390 ha (Fig. 9).

The area available for subtidal culture (0 - 5 m) is 26 596 ha, most of which is in the upper gulf and elsewhere close to the shoreline (Fig. 9). In subtidal areas oysters could be cultured on stakes, platforms or other raised structures, with the height determined by the elevation within the subtidal area.

In waters deeper than 5 m oysters and mussels can be farmed by suspending them from rafts or ropes, and fishes can be cultured in cages. The remaining area of the inner gulf, 42 622 ha, is in this depth category (Fig. 9).

Figure 9. Locations and surface areas available for intertidal, subtidal and suspended culture in the upper inner gulf area. The distribution of mangroves also is shown.

3.4.1 Culture in sheltered areas

Protection from N winds, the strongest in the gulf area, and S winds, the most persistent, is important in several ways. Wind-generated water currents can raise sediments into the water column which interfere with the respiration of molluscs and fish. Further, in intertidal and shallow sub-tidal areas, wave turbulence may damage oysters or mussels directly, the structures on which they are being cultivated, or interfere with spatfall. In deeper waters culture installations could be damaged by storm waves, or maintainance costs could be high due to wear and tear from wave motion.

Optimizing the locations for aquaculture development by providing a 3.7 km (2 nm) lee from both N and S winds leaves a sheltered intertidal area of 946 ha, a sheltered subtidal area of 5 274 ha, and a sheltered openwater area of 3 707 ha (Fig. 10). The surface areas available for other bathymetric-shelter combinations are shown in Table 5.

Table 5

Culture opportunities in sheltered areas in the inner gulf
(areas in hectares)

Type of culture	Gross Area	Sheltered from winds N	S	N & S
Intertidal (> 0 m)	3 390	1 671	728	946
Subtidal (0 - 5 m)	26 596	8 071	7 186	5 274
Cage (> 5 m)	42 622	8 705	10 690	3 707

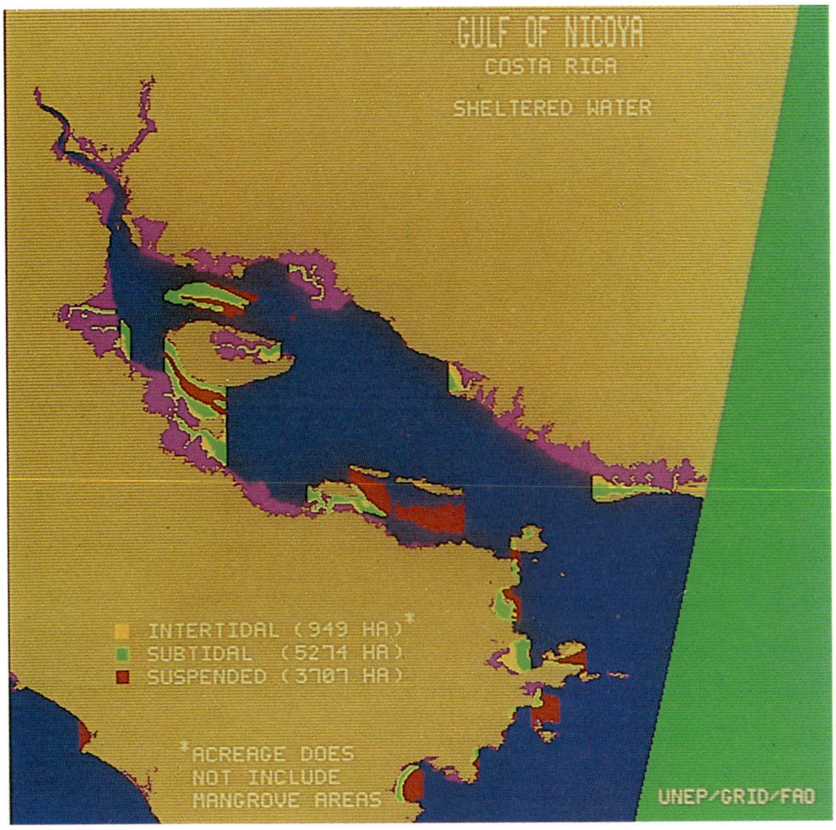

Figure 10. Locations and surface areas for the development of intertidal, subtidal and suspended cultures based on bathymetry and which are also within a 2 nm (3.7 km) lee of N and S winds. Purple indicates mangroves.

3.4.2 Culture in mangrove areas

There are 11 837 ha of mangroves along the gulf periphery, and mangroves make up 65 percent (320 km) of the inner gulf shoreline as determined by analysis of the 15 April 1985 Landsat TM image. Waterways in mangroves offer opportunities for the development of fish and mollusc culture compatible with other uses such as fisheries and tourism. Advantages of culture in mangroves have been discussed by Kapetsky (1985) among which are shelter, availability of seedstock, high biological production, and inexpensive site acquisition. In Cuba the mangrove-water interface is used for the culture of oysters on branches obtained from mangroves (Ministerio de la Agricultura y Representación de la FAO en Cuba, 1984). In the gulf, nearly all of the solar salt ponds have been developed in mangrove areas.

3.4.3 Salinity and water quality

The most extensive areas for subtidal culture occur north and south of Chira island, but are likely to have relatively low salinities during the rainy season and annual salinities which vary widely. There is an elongated area south of Chira Island with depths sufficient for suspended culture, as well (Fig. 10). These locations would require euryhaline organisms. From the point of view of water quality, the locations S of Chira Island might be favoured because they would be under the influence of the Morote River drainage which is little developed for agriculture. The locations N of Chira Island may be influenced by the outflows of the Tempisque and Bebedero drainages, thought to be likely to have a water quality relatively low due to agricultural development. Large, well-sheltered areas for suspended culture are on the south sides of Bejuco and Caballo Islands (Fig. 1). There, salinities would be much higher (see Figures 4, 5 c and 5d for an indication) than further up the inner gulf, and as there are no rivers nearby, water quality would probably be quite good. In contrast, the expanse of subtidal area in the vicinity of Puntarenas could be undevelopable because of domestic sewage from the city.

3.5 Opportunities for Shrimp and Fish Farming in Solar Salt Ponds

In order to locate the solar salt ponds which would be optimum for extensive shrimp and fish culture, a number of factors were taken into account. First, we assumed that shrimp would be the priority crop and therefore that the ponds should be selected to maximize shrimp yield rather than fish production. Second, it was necessary to provide an up-to-date inventory by locating each salt "farm" pond and estimating its surface area.

Apart from management, the success of extensive shrimp farming in salt ponds will depend on the availability of shrimp post-larvae for tidal self-stocking, if supplemental stocking is not pacticed, and proximity of the farms to processing and marketing facilities.

3.5.1 Pond inventory and size characteristics

The total number of saltworks ponds on the gulf shoreline 1 ha and larger is 80 and the cumulative pond surface area is 656 ha. Ponds ranged in size up to 74 ha, but the modal size was 2 ha and more than one-half of the ponds were 5 ha or less (Fig. 11). The locations of ponds is summarized with regard to the mangrove zones in which they occur in Figure 12 which is keyed to Table 6. Annex 2 provides the Universal Transverse Mercator coordinates, latitudes and longitudes and surface areas of the individual ponds. The surface areas indicated may include some salt ponds which are unsuitable for shrimp culture because they are too shallow, too high in elevation to promote circulation, or too salty during the dry season. These are the ponds, usually small, which are used in the final stages of salt concentration.

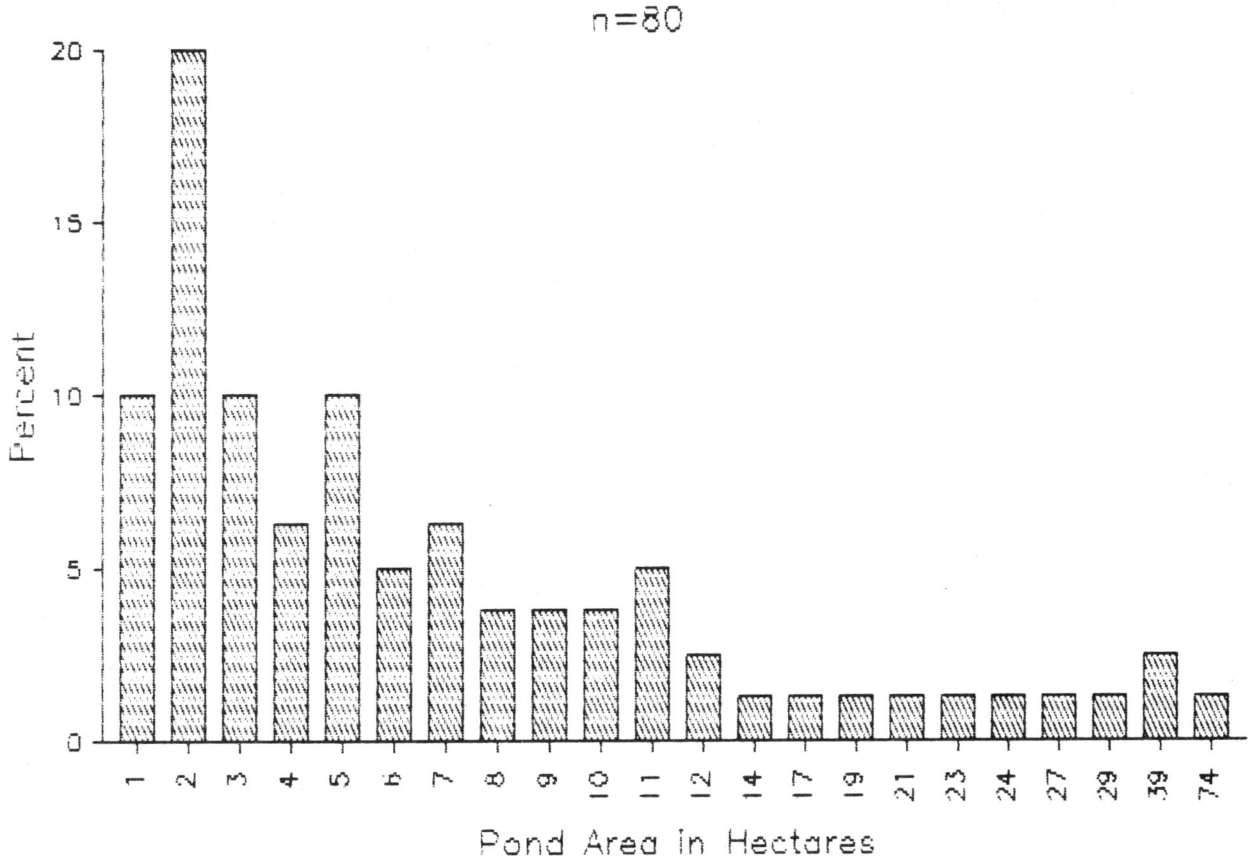

Figure 11. Size distribution of 80 solar salt ponds larger than 1 ha identified by spectral analysis of the Landsat Thematic Mapper image of 15 April 1985. Note that the X-axis is not continuous.

Table 6

Locations and surface areas of solar salt ponds in relation to the mangrove zones portrayed in Figure 12; Shoreline complexity of mangrove zones

Mangrove Zone Number (Fig. 12)	Shoreline Complexity	Solar Salt Ponds Number	Solar Salt Ponds Surface Area (ha)
1		13	80
3		4	11
4	Low	8	69
5		7	65
11		9	85
6		2	14
7	Medium	8	40
10		12	62
8		6	74
12	High	7	135
13		4	21
TOTALS		80	656

3.5.2 Salinity and availability of shrimp post-larva for stocking ponds

In addition to current velocity patterns, the availability of shrimp post-larvae for stocking solar salt ponds is influenced by salinity and habitat.

Salinity affects the selection of sites for extensive shrimp farming in solar salt ponds in two ways. The first is the apparent different sensitivities of various shrimp species to salinity which affects spatial and seasonal distribution of post-larvae. The second is the effect of salinity on growth. For the latter, Penaeus vannamei, one of the most desirable species for culture, does best at salinities ranging from 15 to 20 ppt. Miget (1985) assess growth - salinity relationships representative of most penaeid shrimp species in the following way:

Salinity (ppt)	Growth
< 5; > 40	Poor
5 - 15; 30 - 40	Fair
15 - 30	Good

Attempts to culture shrimp in other than optimum salinity regimes will result in somewhat lowered growth and production, and ultimately, in less economic returns.

Here we assume that the Gulf of Nicoya situation is similar to that of the Pacific coast of Panama and Mexico, where shrimp larvae vary by species and in abundance according to the season. Hughes (1985) notes that in Panama Penaeus vannamei juveniles and post-larvae are found in low salinity areas, P. stylirostris in intermediate salinities, and P. occidentalis in higher salinities. During the dry season P. stylirostris seed is most abundant, and P. vannamei becomes predominant as the rainy season begins and reaches its highest abundance when rains are heaviest. Edwards (1978) found that about 90 percent of the wet season post-larval recruitment to a Mexican Pacific coast lagoon system was P. vannamei. When lagoon salinities rise above 20 ppt, P. vannamei recruitment falls. Thus, the availability of P. vannamei post-larvae coincides with the season when salt ponds are usually idle, when salinities in the gulf are lowest due to direct rainfall and freshwater inflow from rivers.

Based on the observations in Mexico and Panama cited above, we predict that P. vannamei post-larvae will be most abundant for the longest time in the low-salinity areas of the gulf. Thus, solar salt ponds in the upper inner gulf and elsewhere near river mouths would be the prime locations to take advantage of natural concentrations of P. vannamei for self-stocking of ponds.

Based on the complexity of the mangrove-water interface, the locations where shrimp post-larvae are expected to be most abundant are in Mangrove Zones 8, 12, and 13 (Table 6; Fig. 12). However, of these three zones, two, 12 and 13, probably have relatively high salinities the year around because of the adjacent arid Nicoya Peninsula, but Mangrove Zone 8 may have lower salinities due to the influence of the inflow of the Guacimal River (Figs. 1 and 12). Within this zone there are 6 solar salt ponds with a total surface area of 74 ha (Table 6). Mangrove Zones 6, 7, and 10 have medium shoreline complexity. Of these, Zones 6 and 10 are in the low-salinity upper inner gulf area. Thus, solar salt ponds in these two zones, of which there are 14 with a total surface area of 76 ha, could benefit both from salinity conditions favourable for P. vannamei and from a medium complexity post-larval habitat.

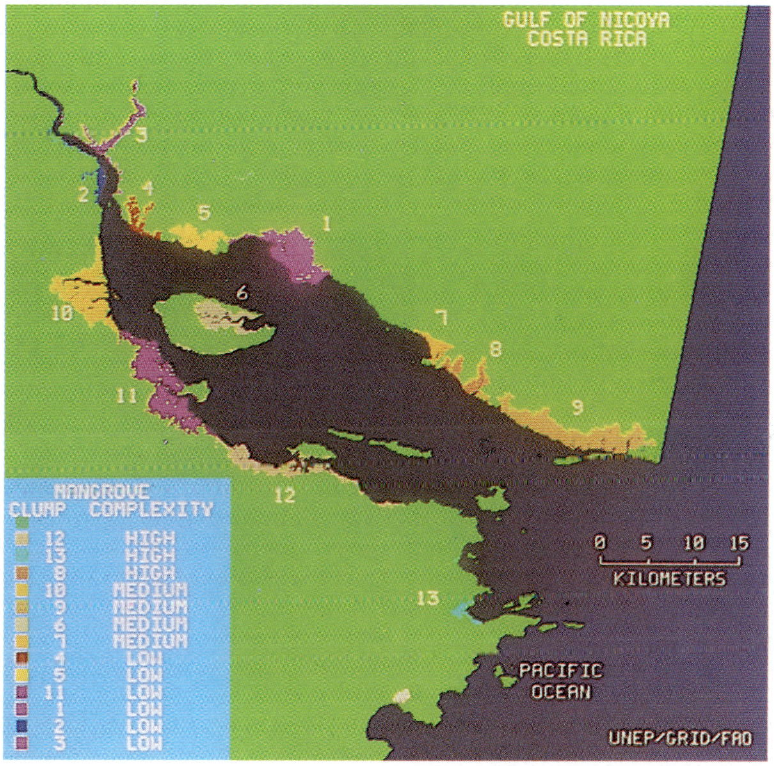

Figure 12. The mangrove zones of the inner gulf are rated according the complexity of the mangrove-water interface. The locations of solar salt ponds are keyed to the numbered areas.

3.5.3 Solar salt ponds in relation to infrastructure

Of the zones predicted to be most favourable for the development of extensive shrimp farming in solar salt ponds on the basis of salinity and shoreline complexity, Zones 8, 10 and 6, Zone 8 is best situated from the point of view of proximity by road to Puntarenas (64 km) and by having a district centre nearby (Chomes) (Fig. 7). Chomes is also where semi-intensive shrimp farming is already underway. This could favour the development of extensive shrimp farming in solar salt ponds by the nearness of shrimp buyers and transporters as well as the technical skills for pond management. In contrast, the other two zones are distant. Zone 10 is more than 80 km from Puntarenas by road and ferry, and Zone 6 is isolated on Chira Island, about 30 km by water to Puntarenas.

3.5.4 Solar salt ponds in relation to water quality

Water quality is likely to be good at all of the S shore solar salt pond locations. On the N shore, ponds in mangrove zone 9 could be affected by domestic pollution from Puntarenas and to some extent by agricultural contaminants in the upper inner gulf due to Tempisque and Bebedero River outflows.

3.6 Semi-Intensive Shrimp Culture Outside of Mangrove Areas

The basic criteria for the semi-intensive culture of shrimp were that the farms be located away from mangroves, that soils be non-acidic, impermeable and compactible in relative terms, and that sites be no more than 1 km from a perennial river and saltwater. Local water quality, site development and site acquisition costs, and infrastructure also are important to identify optimum areas.

3.6.1 Areas outside of the mangroves

The basic locational constraint for semi-intensive shrimp farming was that it was to be outside of mangrove areas. Experience has shown that the development of shrimp ponds in mangroves can result in acidic conditions which, in turn, lower shrimp growth and survival. Shrimp culture in mangroves can be self-defeating also from the point of view of replacing shrimp nursery areas with shrimp ponds as well as the nursery, feeding and breeding areas of other fished species. Studies have shown that pumped circulation can increase yields to economic advantage in comparison with tide-driven circulation (Gedney, Shang and Cook, 1984). With pumped circulation there is no need for ponds to be at intertide levels. Furthermore, pond construction costs are less outside of the mangrove habitat than inside, and management is facilitated. Finally, there is considerable pressure from environmentalists to conserve mangrove areas in Costa Rica.

As was pointed out above, mangroves occupy 65 percent of the shoreline of the inner gulf. At least part of the remaining shoreline area is steep and in these areas pumping costs to lift saltwater and freshwater to ponds would be high. Thus, in practical terms, it is the back-of-the-mangrove areas where the possibilities for obtaining both freshwater and saltwater are best (Fig. 12).

3.6.2 Locations in close proximity to fresh and salt water

There are five areas around the inner gulf within a 1 km proximity to a river. These are the Morote, Abangares, Lagarto, Guacimal and Aranjuez Rivers. The surface area available at these 5 locations, which also are within 1 km of saltwater, is 2 491 ha. Figure 13 shows the merging of the freshwater and saltwater proximities to define 3 of the 5 locations.

3.6.3 Suitability of soils

Within a 1 km distance of the shoreline only 1 percent of the area has soils rating relatively high in their suitability for shrimp ponds and only 27 percent of the area has soils of the second rank (Table 7). In fact, much of the gulf shoreline, particularly that on the S side of the gulf, is made up of the lower ranking soils (Fig. 14) whereas those on the N shore fall into the intermediate or higher ranks. For shrimp farming, the potentially most dangerous soil is the Typic Sulfaquent which makes up 27 percent of the shoreline. Areas of the higher-ranking soils are near the mouth of the Tempisque River and in the mid-gulf area on the N shore.

The site of the present shrimp farm at Chomes (Fig. 7) would not have been selected according to the soil ratings used herein because all of the area is shown as Unit E6 (Table 7; Fig. 14).

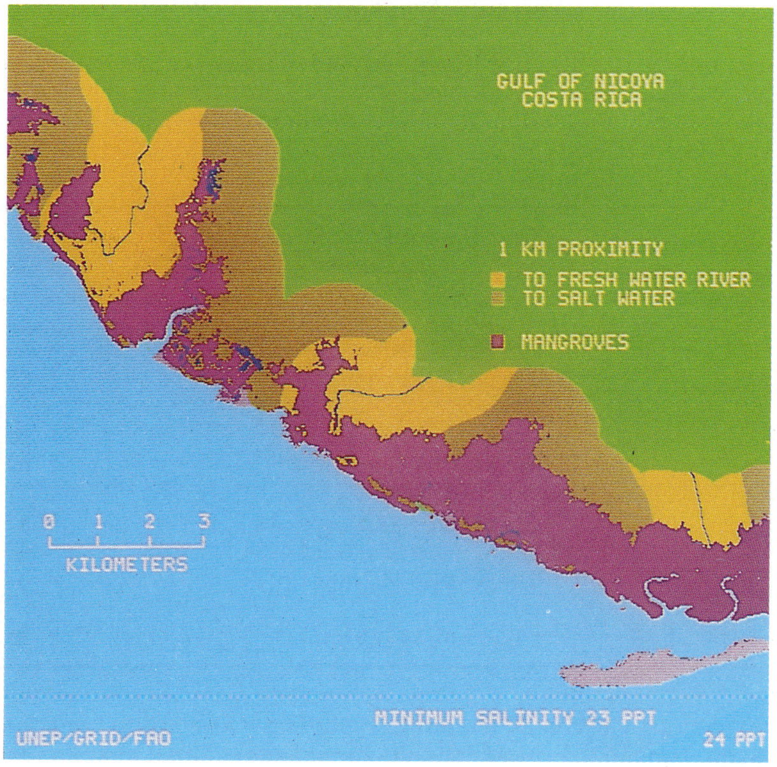

Figure 13. Areas within 1 km of a river (yellow bands) and wihin 1 km of saltwater (beige bands) which are also outside of mangroves. The areas of intersection of the beige and yellow bands define 3 possible locations for seimi-intensive shrimp farming along the N shore of the inner gulf based on proximity to fresh and saltwater. The Puntarenas peninsula is at the lower right side.

Figure 14. Relative suitability of soils for shrimp ponds in a band within 1 km of the inner gulf shoreline

Table 7

Soil units within a 1 km distance of the gulf shoreline

Unit ID.	Rank (relative)	Unit Names	Area (ha)	Percent by rank
MS	1	Fluvaquentic Hapludoll	0	
I24	1	Aquic Ustropept	183	1
I21	2	Fluventic Ustropept	1 539	
I22	2	Fluventic Ustic Dystropepts	2 613	27
I33	2	Ultic Dystropept	1 575	
I19	3	Typic Ustropept	1 086	17
I23	3	Lithic Ustropept	2 617	
V1	4	Vertisols (Grumosols)	4 666	
E5	4	Lithic Ustorthent	939	28
I32	4	Ustic Dystropept	384	
E6	5	Typic Sulfaquent	5 880	27

Seven soil types occur on the 5 potential shrimp farm locations that are within 1 km of salt and fresh water (Table 8). Of the 5 locations, the Lagarto area has the largest proportion of better soils and the Morote site has generally the lowest ranking soils. At 3 of the potential sites shown here for comparison (Fig. 15), nearly all of the Aranjuez site is on the lowest ranking soil type, as is much of the Guacimal area while Lagarto has a relatively large proportion of high ranking soils.

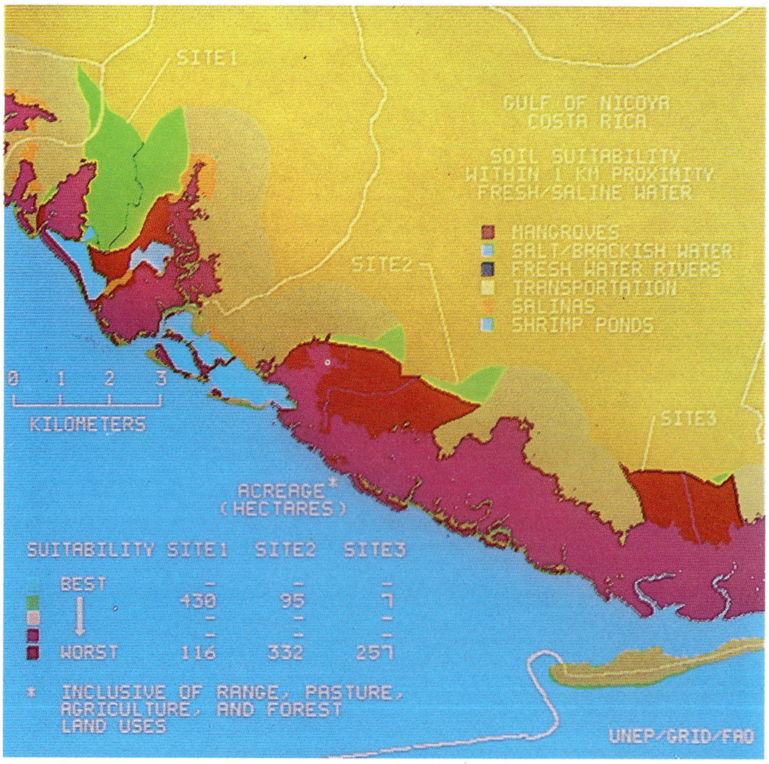

Figure 15. Relative ratings of soils at 3 potential semi-intensive shrimp farming locations along the N shore of the inner gulf. The 3 locations previously were defined by proximity to fresh and saltwater and by being outside of mangroves. Land uses nearby and the road-transportation network are also shown.

Table 8
Types and areal expanses of soils at 5 potential shrimp farm sites within 1 km distances of fresh and salt water. The soils underlie forest, range, pasture and crop land uses, but do not include existing shrimp or solar saltponds

Unit ID.	Rank (relative)	Unit Names	Sites (Areas in ha) 1 Lag	2 Gua	3 Ara	4 Mar	5 Aba
I21	2	Fluventic Ustropept					47
I22	2	Fluventic Ustic Dystropepts	430	95	7		
I19	3	Typic Ustropept					125
I23	3	Lithic Ustropept					26
V1	4	Vertisols (Grumosols)				447	
I32	4	Ustic Dystropept				139	
E6	5	Typic Sulfaquent	116	332	257	33	178
Totals			546	427	264	619	376

1: Lagarto; 2: Guacimal; 3: Aranjuez; 4: Morote; 5: Abangares

3.6.4 Land use determinants of shrimp farm sites

Land uses which were indicative of the suitability of shrimp farm sites were interpreted in three ways: as suggesting local water quality, in terms of site acquisition costs, and as indicative of site development costs.

Table 9 indicates the land uses as interpreted from the satellite image in the areas within 1 km of freshwater and brackishwater. These are portrayed for three of the sites for illustration in Fig. 16.

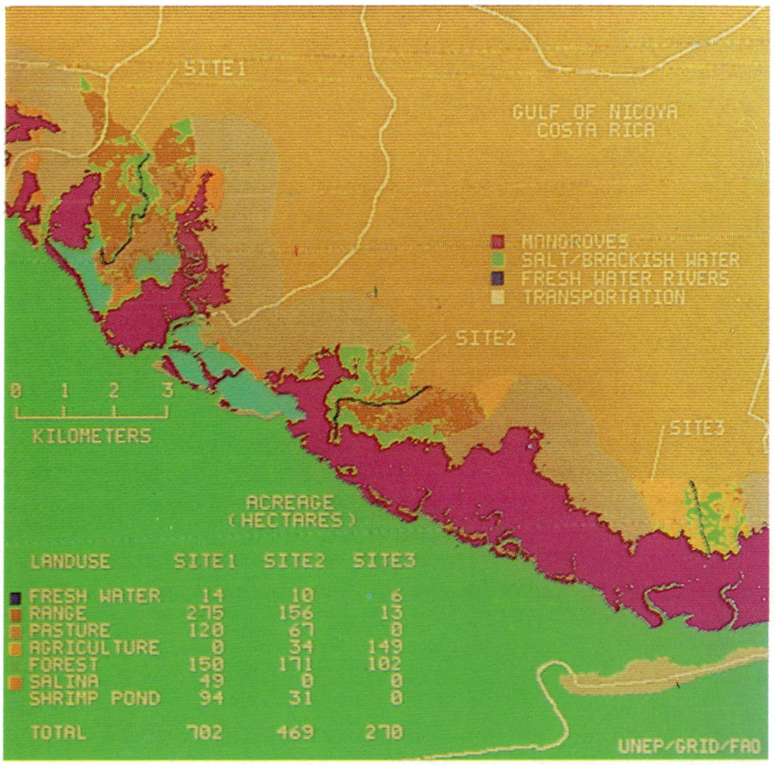

Figure 16. Land use inventories at 3 potential semi-intensive shrimp farming locations along the N shore of the inner gulf based on satellite remote sensing. The locations were originally defined by proximities to water and by being outside of mangroves. Soil types underlying these locations already have been enumerated and rated as to their suitability for shrimp ponds.

Table 9

Land uses on 5 potential sites for semi-intensive shrimp farming within 1 km of fresh and saltwater

Land Use (ha)	Potential Sites (named for rivers) Lagarto (1)	Guacimal (2)	Aranjuez (3)	Morote (4)	Abangares (5)
Freshwater	14	10	6	16	10
Range	275	156	13	125	265
Pasture	120	67	0	236	15
Agriculture	0	34	149	0	0
Forest	150	171	102	258	98
Solar salt ponds	49	0	0	3	26
Shrimp pond	94	31	0	0	0
Totals	702	469	270	638	412

From the point of view of local water quality the Aranjuez area could have problems because it has a large proportion of agriculture, amounting to 55 percent of the area. Water quality problems could arise from pesticides used on agricultural crops. In contrast, use of water eutrophicated by agricultural runoff could enhance aquaculture productivity. More detailed investigations of the kinds and expanses of individual crops would be required to gauge the impact of local agriculture on aquaculture development. This could be done by acquiring satellite imgery during the time when crops are still in the fields so that identification and quantification of individual crop types could be made.

From the standpoint of land acquisition costs (assuming all of the area at each location were to be acquired), the Abangares, Morote and Lagarto areas probably would be cheapest because they have the highest proportions of forest, pasture and range and no relatively expensive land in agriculture.

Site development costs would be relatively low on cropland because there would be no trees to clear, somewhat higher on pasture because of trees along fence lines, and higher again on range land because of trees and shrubs interspersed with less substantial growth. Forest land is the most expensive to clear because of the need to remove the many trunks, roots and underbrush. In this regard, Guacimal, Aranjuez, and Morote would be relatively expensive locations to prepare because of high proportions of forest, but of the three, Aranjuez also has a high proportion of agriculture land, and thus land preparation costs would be low compared to range and pasture land.

Obviously, much data on site acquisition and site development costs would have to be assembled to enable a better informed selection of sites.

3.6.5 Availability of shrimp post-larvae

Close proximity of shrimp post-larvae for supplemental stocking could be an advantage until hatcheries are in operation. According to the results of the mangrove shoreline complexity analysis and salinity, relatively high densities of post-larval *P. vannamei* are predicted for Mangrove Zone 8 (high complexity) which defines the Guacimal site, and for Mangrove Zones 10, (Morote site), 7 (Lagarto) and 9 (Aranjuez), all of which are medium complexity. The Abangares site is in a relatively low-salinity area thereby favouring the presence of *P. vannamei*, but the habitat complexity is low.

3.6.6 Infrastructure

The four N shore sites have the advantage over the S shore site in road transport time and distance to Puntarenas. Morote, (Site 5), is about 90 km, Abangares (Site 4), about 85 km, Lagarto (Site 3) 65 km, Guacimal (Site 2), 64 km, and Aranjuez, (Site 1) about 30 km (Fig. 7; Table 4).

Each site has one or more villages nearby to provide labour and local supplies.

All of the N shore sites have electricity nearby, but the Morote site may not have electrical service.

3.7 Summary of Aquaculture Opportunities

From the point of view of infrastructure the N shore of the gulf is most favoured for aquaculture because its road network is uninterrupted by ferry links, much of it has electrical service, and it is in relatively close proximity to Puntarenas and San José.

Euryhaline organisms can be cultured in the salinity regime of the upper inner gulf and at sites near the mouths of rivers. Opportunities for the culture of organisms less tolerant of low salinities, or of salinity variations, occur in the lower inner gulf open-waters and along the S shore.

Water quality is probably best along the S shore of the inner gulf, in the vicinity of the smaller islands and near Chira Island. The upper inner gulf could present problems from agricultural pesticides and from high turbidities associated with the Tempisque-Bebedero drainages.

3.7.1 Intertidal, subtidal and suspended cultures

The inner gulf occupies an area of about 72 600 ha of which only about 14 percent is optimum for intertidal, subtidal or suspended cultures. The optimum areas for the development of subtidal and suspended cultures when shelter from N and S winds and security are considered are mainly in the vicinity of Chira Island. Altogether there are 5 274 ha of subtidal area which are well protected (20 percent of the total subtidal area), and 3 707 ha well sheltered for suspended cultures (only 9 percent of the area 5 m or greater in depth). The potential sites in the vicinity of Chira Island are lower salinity areas during the rainy season because of the influence of the Morote, Abangares and Tempisque Rivers. Therefore, this area lends itself to the culture euryhaline organisms. The locations south of Chira Island are relatively distant from Puntarenas whereas those north of Chira Island might be more easily developed by taking advantage of infrastructure on the north shore of the gulf. Large areas sufficiently deep for suspended culture also exist south of Bejuco and Caballo islands where the salinity is higher and the sites are in close proximity of Puntarenas by water. Well-sheltered intertidal areas are relatively few around the gulf shoreline, amounting to only 946 ha, about 28 percent of the total intertidal area. There are several areas in the vicinity of Berugate Island, and another near Pta Morales. The intertidal area around Puntarenas is relatively large, but may be polluted.

3.7.2 Extensive aquaculture in solar salt ponds

Eighty solar salt ponds with a total surface area of 656 ha were detected using satellite remote sensing. Those best suited to be developed for extensive shrimp culture will be in areas where *Penaeus vannamei* post-larvae are most abundant, where pond salinities can be manipulated to achieve suitable ranges, and which are supported by good road or water communications with Puntarenas.

Optimizing locations for *P. vannamei* by favouring moderate salinities for grow-out indicates locations in the upper inner gulf under the influence of Tempisque outflows, and elsewhere near the N shore river mouths and the Morote River on the S shore. Optimizing solar salt pond locations according to post-larval shrimp habitat (i.e., mangrove-shoreline interface along with a salinity regime favourable for *P. vannamei*) suggests three locations with a total of 20 ponds and 150 ha of surface area, equivalent to 25 percent in number and 23 percent of the available solar salt pond surface. The salt ponds near Chomes have the best infrastructure including proximity to exisiting shrimp farms.

3.7.3 Semi-intensive shrimp farming outside of mangroves

Five locations around the inner gulf totaling 2 232 ha, and of which 4 are on the N shore and 1 on the S shore, are optimum for semi-intensive shrimp farming based on the availability of fresh and salt water nearby. Information on soil suitability and land uses on each of the five sites enables further consideration in terms of local water quality, site preparation and site acquisition costs. Availability of infrastructure, particularly road communications and electricity, favours the N shore sites.

4. DISCUSSION AND CONCLUSIONS

The present study has demonstrated that data on a number of environmental and economic parameters can be compiled from diverse sources, merged in a Geographical Information System and interpreted to evaluate opportunities for aquaculture development in terms of the most promising locations and the amounts of land and water available.

4.1 Accuracy of the Results

It is important to keep in mind that the results are indicative of opportunities for aquaculture development and are for general planning purposes. Investigations on the ground and in the water are necessary to verify the suitability of individual areas identified herein before development programmes can be planned in detail.

There are accuracy limits on the results, just as in any other kind of study. The maps and tabular data used to compile the GIS have come from a variety of sources based on data not all of which is recent and not all of which has the same amount of detail and locational accuracy. For example, the bathymetry is based on a nautical chart at 1:100 000 scale (1 km = 10 mm), soils were mapped at 1:200 000 scale (1 km = 5 mm), and much of the transportation network was taken from a map at 1:500 000 (1 km = 2 mm).

Some caution has to be used in the interpretation of GIS results. Soils provide an example. Soils are not spatially homogeneous in the first place, even if well-mapped at a large scale. Soils were compiled for the GIS as mapped at 1:200 000. Therefore, the actual locations of the soil boundaries underlying the 3 potential shrimp farm sites identified in Figure 13 is somewhat uncertain although this is not apparent from the map itself. Soils mapped at 1: 20 000 would have been more appropriate for merging with the land uses obtained from satellite imagery. With such detail of information a GIS could be compiled for site selection.

Burrough (1986) provides a detailed discussion of data quality, errors and natural variation associated with Geographcal Information Systems.

4.2 Supplementary Ground and Water Investigations

Clearly, the larger the number of parameters evaluated, the more comprehensive the assessment of aquaculture development potential. In this regard there are a number of parameters which could be added to the results of the present study to better pinpoint areas with development possibilities. Also, additional data on parameters already analyzed could improve evaluations. These are listed below:

Infrastructure

- Update map of electrical services and redraw to a larger scale; determine reliability of electrical services

- Obtain village and district center population sizes to improve knowledge of labour pool

- Review national and local development plans which could affect the gulf area in the near future(e.g., planned road construction; export market promotion)

"Ecology"

- Improve spatial distribution of salinity data to include the shoreline areas already identified as being optimal sites for aquaculture development

- Include actual data on post-larval shrimp abundance in time and spatially to improve the predictions made herein

- Measure current velocities to establish seasonal current patterns at sites already identified as promising for culture of molluscs and fish, and to better assess larval shrimp availability

- Verify dry season flows of the Guacimal and Aranjuez Rivers

- Research the distribution and depth of groundwater to determine if freshwater and saline layers could be tapped for semi-intensive shrimp farming to enable siting of shrimp farms away from rivers and mangroves

- Map bottom types in the intertidal and subtidal areas already identified as being advantageous for mollusc culture

- Use satellite remote sensing to determine the temporal and spatial distribution of suspended sediments and chlorophyll to improve siting for mollusc culture

- Incorporate topography as an additional site selection criterion for semi-intensive shrimp farming outside of mangroves

Water Quality

- Merge April 15 1985 Landsat TM image with March 23 1985 image (which is adjacent) to cover all the drainage basins of the gulf, or acquire more recent images

Land uses

- Acquire early dry season (or late wet season) satellite image to identify each crop type; this, in turn, would provide better information for inferences on dangers from pesticide pollution and on site acquisition and development costs

Economics

- Estimate costs of land acquisition for various kinds of croplands, rangelands, pasture and forest at places already identified as promising for development

- Estimate cost factors based on travel time and distance to establish operating costs for road and water transport according to location

4.3 The Utility of GIS's and Remote Sensing to Plan for Aquaculture Development

The results of this study clearly show that GIS and satellite remote sensing technologies can be combined to provide a variety of information useful for general planning for aquaculture development as well as to identify and rank individual locations. Put another way, the application of these technologies makes it possible to quickly identify the locations which are marginal or outright unsuitable for aquaculture development. Thus, human and financial resources can be focused on areas which are apparently optimum for further, more detailed investigations.

Although this approach is technologically feasible, a question equally important is whether or not it is cost effective, particularly in the context of developing countries. In comparison with manual methods, computer automation makes the work time-effective, not so much in the assembly of information to make individual maps, but rather in analyses (merging data from several maps, and in extracting tabular information from them such as surface areas and distances) and the map-updating process itself. The computer-automated approach also provides flexibility. New data, or modifications of exisitng data, can be added and manipulated very quickly to produce new maps, graphics, or tables. Also, alternatives ("what if" questions) can be rapidly evaluated. For example, "What areas remain for subtidal mollusc culture if shelter is reduced from 2 nm to 1 nm?"

This study began in July, 1986 with preliminary analyses of satellite data to assist ground verifications based on a plan of work conceived earlier. Ground truthing and local data collection were conducted in August, and a detailed work plan for subsequent analyses was drawn up immediately thereafter. The first results were available in October, the second set of results in February 1987, and the final results in March and April. Reporting was completed in April 1987. However, this has been a part-time activity for all persons involved. We believe that the study could have been completed in about two months of full-time work for two persons.

As suggested above by the number of times the results were reported, this work is interactive. Not all of the outcomes can be anticipated until the analyses are complete. The results of the preliminary analyses suggest additional avenues for investigation. If these are pursued, such as in the present situation, then additional time is required to complete the additional analyses and reporting.

An important question about the efficiency of the GIS-remote sensing approach to planning for aquaculture in developing countries is the availability of personnel with the necessary skills. The persons and skills involved with this study included a fishery biologist with specialized knowledge of aquaculture development in terms of environmental and infrastructural needs and a general knowledge of remote sensing and GIS applications. A second person was a systems analyst specialized in satellite data image processing and GIS. An important participant was a fisheries and aquaculture administrator from Costa Rica with a technical background in aquaculture, training in remote sensing, familiarity with the study area, and knowledge of the locations of much of the map and tabular data incorporated in the GIS. The water quality study using remote sensing (Annex 1) required a limnologist with a good working knowledge of remote sensing applications for the study of the aquatic environment.

The GIS was compiled on sophisticated and costly computer hardware and software which is out of the financial reach of most fishery departments in developing countries. There are several solutions to this problem. One is for fishery and aquaculture departments to share time and expenses for the development of a central GIS facility with other departments of government. In favour of this strategy is that much of the information required for fisheries and aquaculture development is also required for development planning in other fields such as agriculture and hydrology. Thus, data acquisition and data analysis costs (such as for satellite data) can be shared. Another advantage is that the costs of employing a highly specialized systems analyst for image processing and GIS compilation is shared as are costs for supporting personnel. FAO has adopted a central GIS approach to serve the similar personnel and equipment needs but varying technical and information requirements of each of its departments.

An alternative solution is to purchase a microcomputer-driven image processing and GIS system. The advantages are that such systems are relatively inexpensive, say US$ 50-60 000 for software, hardware and peripherals of the best quality, and they are easy to learn to use. Furthermore, with the hardware and software to hand, updating of current work and new investigations can be undertaken when required. The disadvantage is that analyses are slow, and the variety of analytical techniques is not so broad as with larger systems. Furthermore, data acquisition costs will be relatively high, if they are not shared among users from other disciplines, and some personnel have to be allocated to manage the analyses.

Satellite remote sensing proved to be a useful tool to provide data for the GIS for this study. In general, where other sources of data are spatially limited or out of date, then remote sensing can be a cost-effective alternative well worth considering for many applications. In the present study we were able to make inferences about water quality in river and island drainages and to indicate relative site development and acquisition costs using various interpretations of remotely determined land uses as surrogates. Based on the same imagery and previous satellite studies (Annex 1), some parameters of the quality of gulf water were directly measured.

New satellite sensors especially designed to measure water features, and an increasing number of national remote sensing centres favour the more frequent use of remote sensing for aquaculture applications in developing countries in the near future. Constraints in some areas are cloud cover which precludes synoptic coverage, especially during the dry season, and the cost of imagery (e.g., US$ 3 500 for a Landsat Thematic Mapper computer compatible tape (CCT), such as used in this study, each scene covering about 32 000 km^2).

A solution to cloud cover is to substitute airborne photography for satellite data when spectral analyses are not required. Recent aerial photography (black and white) is often available from national geographic institutes at very low cost. Costs of satellite data can be reduced by sharing image purchase and processing costs among users, as suggested above. Another cost-saving possibility is to substitute weather satellite (Advanced Very High Resolution Radiometer) images for Landsat or SPOT coverage in instances where synoptic coverage is important to undertstand water dynamics, where the area under study is relatively large, and where resolution (1.1 km for AVHRR at best) is less important. AVHRR CCT cost several hundred dollars each and the coverage is twice daily. Present capabilities include surface temperature and turbidity, but chlorophyll estimation may be developed.

In summary, we believe that Geographical Information Systems can be useful tools for aquaculture planning. The utility of aquaculture GIS's can be most fully realized by integration and analysis of aquaculture productivity data, economic information on land and water resources, economics of aquaculture operations and marketing along with the mainly environmental parameters which were the focus of this study. In this way a GIS could be used to assess aquaculture development potential very comprehensively.

5. ACKNOWLEDGEMENTS

The final version of this document benefitted from reviews of an earlier version by H.F. Henderson and M. Pedini. J. Lawrence-Amendola processed the test and figures, G.A.Soave checked the literature references and A. Meschini drafted the figures.

6. LITERATURE CITED

American Farmland Trust, 1985. Survey of geographic information systems for natural resource decision making at the local level. Washington, D.C., American Farmland Trust, 25 p. + 5 app.

Amos, C.L. and T.T. Alfoldi, 1979. The determination of suspended sediment concentration in a macrotidal system using Landsat data. J.Sediment. Petrol., 49(1):159-74

Angell, C.L., 1986. The biology and culture of tropical oysters. ICLARM Stud.Rev., 13:42 p.

Bartels, C.E. et al., 1983. Occurrence, distribution, abundance and diversity of fishes in the Gulf of Nicoya, Costa Rica. Rev.Biol. Trop., 31(1):75-101

____________, 1984. Ecological assessment of finfishes as indicators of habitats in the Gulf of Nicoya, Costa Rica. Hydrobiologica, 112:197-207

Bowker, D.E. et al., 1975. An investigation of the waters in the lower Chesapeake bay area. Proc.Int.Symp.Remote Sensing Environ., 10:411-20

Burrough, P.A., 1986. Principles of geographical information systems for land resources assessments. Oxford, Clarendon Press, Monographs on soil and resources survey, 12:193 p.

Cheney, D.P. and H.R. Rabanal, 1984. Remote sensing and its application to inland fisheries and aquaculture. FAO Fish.Circ., (768):50 p.

Clifford, H.C. III, 1985. Semi-intensive shrimp farming. In Shrimp farming manual, edited by G.W. Chamberlain, M.G. Haby and R.J. Miget. Corpus Christi, Texas, Texas Agricultural Extension Service, IV-15 to IV-42

Commission for Inland Fisheries of Latin America, 1987. Report of the second session of the Working Party on aquaculture. Guayaquil, Ecuador, 22-26 September 1986. FAO Fish.Rep., (373):36 p. Issued also in Spanish

Conant, F. et al., 1983. Resource inventory and baseline study methods for developing countries. Washington, D.C., American Association for the Advancement of Science, 83-3:539 p.

Dean, H.K. et al., 1986. Trace metal concentrations in sediment and invertebrates from the Gulf of Nicoya, Costa Rica. Mar. Pollut.Bull., 17(3):128-31

Dittel, A.I., C.E. Epifanio and J. Bautistata Chavarria, 1985. Population biology of the Portunid crab Callinectes arcuatus Ordway in the Gulf of Nicoya, Costa Rica, Central America. Estaur.Coast.Shelf Sci., 20:593-602

Dow, D., 1982. Software programs to measure interface complexity with remote sensing data, with an example of a marine ecosystem application. Washington, D.C., National Aereonautics and Space Administration, Earth Resources Laboratory, report (210):25 p.

Edwards, R.R.C., 1977. Field experiments on growth and mortality of Penaeus vannamei in a Mexican coastal lagoon complex. Estuar.Coast.Mar.Sci., 5:107-21

__________, 1978. The fishery and fisheries biology of penaeid shrimp on the Pacific coast of Mexico. Oceanogr.Mar.Biol., 16:145-80

Epifanio, C.E., D. Maurer and A.I. Dittel, 1983. Seasonal changes in nutrients and dissolved oxygen in the Gulf of Nicoya, a tropical estuary on the Pacific coast of Central America. Hydrobiologia, 101(3):231-8

Gedney, R.H., Y.C. Shang and H.L. Cook, 1984. Comparative study of tidal and pumped water supply for brackishwater aquaculture ponds in Malaysia. Rome, FAO, FI:DP/MAL/77/008, Field document, 2:119-58

Glude, J.B., 1981. The potential for mollusc farming in Costa Rica. A report to the Aquaculture Development and Coordination Programme of FAO. Seattle, Glude Aquaculture Consultants Inc.

Gordon, H.R. and W.R. McCluney, 1975. Estimation of the depth of sunlight penetration in the sea for remote sensing. Appl. Optics, 14(2):413-6

Graham, M.H. et al., 1985. ELAS earth resources laboratory applications software. Vol. I. Programmer reference. Vol. II. User reference. Washington, D.C., NASA National Space Technology Laboratories, Earth Resources Laboratory

Grassl, H. et al., 1986. The use of fluorescence measurements from space for separating constituents of sea water. Summary report. Paris, ESA Contract No. RFQ 3-5059/84/NL/MD:31 p.

Great Britain, Hydrographic Department, 1984. Admiralty tide tables, 1985. Vol. 3. Pacific Ocean and adjacent seas. London, Hydrographer of the Navy, 484 p.

Hargraves, P.E. and R. Viquez, 1981. The dinoflagellate red tide in Golfo de Nicoya, Costa Rica. Rev.Biol.Trop., 29(1):31-8

Hargraves, P.E. and R. Viquez M., 1985. Spatial and temporal distribution of phytoplankton in the Gulf of Nicoya, Costa Rica. Bull.Mar.Sci., 37(2):577-85

Hartshorne, G. et al., 1982. Costa Rica: country environmental profile, a field study. USAID Contract No. 0000-C-00-1004-00. San José, Costa Rica, Tropical Science Centre, p.77, Figure IV-18

Hollin, D., 1985. Preliminary economics of shrimp mariculture in Texas. In Shrimp farming manual, edited by G.W. Chamberlain, M.G. Haby and R.J. Miget. Corpus Christi, Texas, Texas Agricultural Extension Service, VII-1 to VII-37

Hughes, D.G., 1985. Aquaculture: Central America. In Proceedings. Shrimp world market conference. Acapulco, Mexico, November 30-December 2 1984. New Orleans, Louisiana, Shrimp World Incorporated, pp. 162-6

Instituto Geographico de Costa Rica, 1983. Costa Rica. Mapa Fisico-Político 1:500.000. San José, Instituto Geographico de Costa Rica, Edición 1983

Instituto Geográphico Nacional/Servicio Geodesico Interamericano (IAGS), 1984. Golfo de Nicoya. Washington, D.C., Defense Mapping Agency Hydrographic/Topographic Centre, CR.006 21544

Jerlov, N.G., 1968. Optical oceanography. Amsterdam, Elsevier, 194 p.

Kapetsky, J.M., 1985. Mangroves, fisheries and aquaculture. FAO Fish.Rep./FAO Rapp.-Pêches/FAO Inf.Pesca, (388)Suppl.:17-36

___________, Satellite remote sensing to locate and inventory small water bodies for fisheries management and aquaculture development in Zimbabwe (in press)

Klemas, V. et al., 1983. Water quality assessment of the Golfo de Nicoya, Costa Rica. Progress report of 1981-82 International Sea Grant program. Phase 1 of remote sensing task. Newark, University of Delaware (DEL-SG-04-83):96 p.

Lorenzen, M.W., 1980. Use of chlorophyll-Secchi disk relationships. Limnol. Oceanogr., 25(2):371-2

Loubersac, L., 1985. Study of intertidal zones using simulated SPOT data: inventorying of aquaculture sites in the intertropical zone. In Report on applications of remote sensing to aquaculture and inland fisheries. FAO Remote Sensing Cent.-Ser., (27):261-72

Maurer, D. and J.A. Vargas, 1984. Diversity of soft-bottom benthos in a tropical estuary: Gulf of Nicoya, Costa Rica. Mar.Biol., 81(1):97-106

Miget, R.J., 1985. Coastal aquaculture site selection. In Shrimp farming manual, edited by G.W. Chamberlain, M.G. Haby and R.J. Miget. Corpus Christi, Texas, Texas Agricultural Extension Service, II-1 to II-10

Ministerio de la Agricultura y Representación de la FAO en Cuba, 1984. Informe de Cuba. Seminario "Manejo Integral de Ecosistemas de Manglares". Manzanillo, Cuba, 6-9 Noviembre de 1984. Manzanillo, Ministerio de la Agricultura y Representación de la FAO en Cuba, 23 p.

Mooneyhan, W., 1985. Determining aquaculture development potential via remote sensing and spatial modelling. In Report on applications of remote sensing to aquaculture and inland fisheries. FAO Remote Sensing Cent.Ser., (27):217-48

Munday, J.C., T.T. Alfoldi and C.L. Amos, 1979. Bay of Fundy verification of a system for multidate Landsat measurement of suspended sediment. I. Satellite hydrology. American Water Resources Association, pp. 622-40

Munday, J.C. and P.L. Zubkoff, 1981. Remote sensing of dinoflagellate blooms in a turbid estuary. Photogramm.Eng.Remote Sensing, 47(4):523-31

Perez, S., A. Alvarado and E. Ramirez, 1978. Asociación de subgrupos de suelos de Costa Rica (mapa preliminar). San José, Costa Rica. OPSA/MAG Instituto Geográfico Nacional

Peterson, C.L., 1960. The physical oceanography of the Gulf of Nicoya, Costa Rica, a tropical estuary. Bull.I-ATTC, 4(4):139-90

Phillips, P.C., 1985. Aquaculture as an alternative occupation on the Pacific coast of Costa Rica. ICLARM Newsl., 8(2):14-5

Quayle, D.B., 1980. Tropical oysters: culture and methods. Ottawa, Canada, IDRC (IDRC-TS17e):80 p. Issued also in French

Rodriguez, M.A. et al., 1978. Inventario de las principales cuencas de Costa Rica. Desarrollo Integral de los recursos forestales Costa Rica. San José, PNUD/FAO/COS/72/013, Documento de Trabajo No. 18: 36 p. (mimeo)

Scherz, J.P. and J.F. Van Domelen, 1974. Water quality indicators obtainable from aircraft and Landsat images and their use in classifying lakes. Proc.Int.Symp.-Remote Sensing Environ., 10:447-60

Smith, R.C., J.E. Tyler and C.R. Goldman, 1973. Optical properties and color of lake Tahoe and Crater lake. Limnol.Oceanogr., 18(2):189-99

Travaglia, C. and M. Lorenzini, 1985. Monitoring algae growth by digital analysis of Landsat data: the Orbetello lagoon case study. FAO Remote Sensing Cent.Ser., (31):28 p.

Tyler, J.E., 1968. The Secchi disk. Limnol.Oceanogr., 13(1):1-6

Tyler, J.E. and R.C. Smith, 1970. Measurements of spectral irradiance underwater. New York, Gordon and Breach, 103 p.

United States Army Coastal Engineering Research Center, 1977. Shore Protection Manual. Volume 1. Washington, D.C., U.S. Government Printing Office, pag.var. 3rd ed.

Universidad de Costa Rica, 1985. Report on bottom fauna and sediment composition. Centro de Investigación en Ciencias del Mar y Limnología, Universidad de Costa Rica

Voorhis, A.D. et al., 1983. The estuarine character of the Gulf of Nicoya, an embayment on the Pacific coast of Central America. Hydrobiologia, 99:225-37

Whitley, S.L. et al., 1981. ELAS: a geobased information system that is transferrable to several computers. Proc.Int.Symp.Remote Sensing Environ., 15 Vol.3:1091-8

Annex 1

Remote Sensing Evaluation of Water Quality in the Gulf of Nicoya (Costa Rica)

by

Jean-Michel Jaquet
Groupe de Télédétection Aquatique, Départment de Géologie, Université
13 rue des Maraichers, CH-1211 Geneva, Switzerland

1. INTRODUCTION

This annex deals exclusively with the problem of water quality evaluation from satellite imagery and ground information. Three Landsat TM and MSS scenes are examined.

To our knowledge the only similar study undertaken in the Gulf is by Klemas et al. (1983)[1]/. They analyzed Landsat MSS imagery in two areas (Rio Tarcoles turbid plume and tidal front around Islas Negritos). They also measured at various dates, not coordinated with satellite overpasses, suspended solid concentration and optical parameters at several stations, mostly in the outer gulf.

We have used their data, as well as others, in an attempt to establish relationships between satellite signature and classical water quality parameters. This endeavour was difficult due not so much to the complexity of the Nicoya estuary, but more to the lack of groundtruthing coordinated with satellite overpasses, and to the small number of available images.

2. GEOGRAPHIC SITUATION

The Gulf of Nicoya's physiographical features are shown in Figure 1[2]/, and depth zones in Figure 9. According to Voorhis et al. (1983), the gulf is a positive estuary characterized by "an east-west asymmetry due to the predominant runoff along its eastern shore from the Barranca and Tarcoles rivers. The freshened surface water from the inner gulf combines with the runoff from these rivers and flows southward along the eastern side of the outer gulf. This flow is compensated by a northward flow of more saline water on the western side at all depths and on the eastern side along the bottom". Water turbidity is governed by this circulation pattern as well as by variations in river runoff and sediment resuspension on shallow banks due to wind or tidal action.

3. SATELLITE IMAGERY

3.1 Available Images

Because of cloud cover, the satellite images represent only dry-season conditions. Their characteristics are listed below:

Type	Date	Place	Tide
TM	15.04.85	Upper Gulf	1.75 m above datum 0.55 m below max. and rising
TM	23.03.85	Lower Gulf	0.5 m above datum 2.5 m below max. and falling
MSS	24.01.84	Lower Gulf	1.25 m above datum 1.45 m below max. and falling

[1]/ The literature cited in this annex is referenced in Section 6 of the main report.

[2]/ Figures not preceded by an "A" are in the main body of the report.

Owing to the area targeted for aquaculture assessment (inner gulf), most of the interpretation was on the April image, with only occasional reference to the others. Computer compatible tapes were processed at GRID by means of the ELAS software. Contrast-stretching was applied to water areas, but "chromaticity" was computed on raw DN values, without prior transformation into radiance.

Some clouds were easily identified in Figure A1, and these areas were discarded in the analysis. Haze is more of a problem, especially over water, where its discrimination from turbidity can be difficult. With the help of TM band 6 (thermal) and chromaticity analysis, haze patches have been delineated and plotted in Figure A4a.

3.2 Coastline and Bathymetry at Overpass Time

Combining the information from the table above and Figure 9 results in the profile drawn in Figure A3. In April, the mud flats are under 0.3 to 1.7 m of water (tide rising), whereas in March they are almost completely emerged (tide falling).

Bathymetry read in Figure 9 has to be incremented by 1.7 m and 0.2 m for April and March overpasses, respectively.

3.3 Bottom Influence

According to Scherz and Van Domelen (1975), sediments located below Secchi depth (Zs) will have no significant influence on the satellite signal. In the inner gulf turbid waters, Zs does not seem to exceed 5 m (Appendix 1), and hence bottom influence on the signal does not go beyond this depth, and is more likely to be less.

In April, the satelilte signal from shallow mud flats was influenced by bottom reflection (Figure A3B) as confirmed by chromaticity analysis. In March (Figure A2) the intertidal zone was almost totally above water, and shows a well-defined signal in the chromaticity diagram (Figure A5).

3.4 Depth of Penetration

A detailed treatment of this problem is given in Appendix 2. On the basis of beam attenuation coefficient and Secchi depth values measured by Klemas *et al.* (1983), one may match Gulf of Nicoya waters to Jerlov's (1968) Type 3 - 9 and use Gordon and McCluney's (1975) Figure 4 to estimate peneration depth (see Table 2 in Appendix 2). Even for maximum Zs (5 m), Z90 barely reaches 4 m in the green band. For true colour images, Z_{90} = 0.85 Zs.

4. WATER FEATURES DESCRIPTION

4.1 April Image

4.1.1 True colour

Visual inspection of Figure A1 reveals the following features summarized on Figure A4:

- Wave patterns in Lower and Middle Gulf, consistent with tidal state (flow). Sea state might be responsible for the occurrence of Light Ocean Water (LOW) in the outer gulf) increased reflectance.

- Widespread planktonic blooms (Figure A4c), either brown or pink (Hargraves and Viquez, 1981), located in the western Lower Gulf and decreasing in intensity northward (faint patch between Caballo and San Lucas Islands). They coexist at places (e.g., Ballena Bay) with dark, wave deformed streaks up to several kilometers long, which could also be planktonic blooms: according to Munday and Zubkoff (1981), equivocal changes in reflectance can result from variations in the depth profile of cell concentration. Another interpretation of these black slicks is the presence of "yellow substance" secreted by the phytoplankton (Grassl *et al.*, 1986).

- Water masses: five broad subdivisions can be drawn based on water colour (Figure A4b). They range from low reflectance ocean water to the highly turbid Tempisque River. Their interpretation is based on chromaticity analysis following Munday *et al.* (1979).

4.1.2 "Chromaticity" analysis

DN and TM bands 2, 3 and 4 have been recombined as X-2/2+3+4 and Y-3/2+3+4. Selected points from various locations in the Gulf are plotted in Figure A5. The information extracted from this diagram is:

- A turbidity gradient between purely riverine water of the Tempisque River and "clean" ocean water with turbidity varying in parallel to (1-X) (Amos and Alfoldi, 1979).

- Domains of similar signal characteristics (Figure A4b) such as:

  * FW: fluvial water mixed with sea water (FWM).

  * EMF: emerged mud flats, close to very turbid water but off the trend (visible only on March image).

  * SMF: submerged mud flats, intermediate between diluted FWM and EMF. The departure from the trend is probably due to the influence of the bottom.

  * IOW: isolated ocean water. Lenses of dark, greenish water around Chira Island, resulting from tidal input of DOW mixed with FW.

  * LOW: "lighter" ocean water, located inshore from DOW masses. It is tentativley interpreted as a more turbid ocean water (resuspension of non-fluviatile sediments).

  * RTW: red-tide water. The signature of red or brown patches is equivalent, in X terms, to turbid river water, but is well off the trend (lower Y), which corresponds to the tendency shown, for chlorophyll, in Munday et al. (1979, page 632).

  * DOW: "dark" ocean water, characterized by maximum values of X and Y (very low turbidity).

  * H: areas affected by haze cannot be distinguished from SMF in X, but are much lower in Y. They overlap with RTW, but discrimination can be achieved through visual inspection of true colour images.

Water mass labels ae superimposed on broader categories in Figure A4b. There is a distinctive pattern of ocean water flooding into the estuary along the south shore, with a mixture of more turbid patches and IOW in the Upper Gulf. Red tides seem to be linked to the presence of DOW. They do not occur in the Upper Gulf.

4.2 March Image

With ebbing tide, turbid water invades the Middle Gulf, leaving only a stream of IOW off Puntarenas. The front at Negritos Islands is clearly seen, as well as the Tarcoles River turbid plume. Pinkish hue at the top centre of the view might indicate a plankton bloom (Figure A2).

4.3 January Image (MSS)

The patterns are similar to those of March, with maximum turbidity being concentrated between Negritos and San Lucas Islands.

5. GROUND-TRUTH DATA

5.1 Bottom-Sediment Composition

Bottom types are indicated at about 20 localities on the nautical charts of the Gulf, most at depths greater than 10 m. In addition, sediment samples were analyzed for grain-size and organic matter content (Universidad de Costa Rica, 1985).

As shown in Figure A6a and b, most of the Inner Gulf bottom consists of green mud, relatively rich in organic matter. The sediments are more sandy and lower in organic matter south of Puntarenas and on the east coast of Nicoya peninsula.

5.2 Salinity

Data collected over several years at many offshore stations (Peterson, 1960) have been used to compute minimal salinity values, contour plotted in Figure 4. The area upstream from Chira Island has salinity values below 10 ppt. At Caballo Island, they do not go below 20 ppt and keep rising down the Gulf. In the absence of inshore stations, salinity patterns represent only general trends.

5.3 Turbidity

No general survey of turbidity in the Gulf being available to date, we had to draw from the report of Klemas et al. (1983), which chiefly covers the outer Gulf. In addition, a Secchi depth profile over the mud banks north of Puntarenas was run in August, 1986 (Kapetsky, written comm.). Owing to the highly variable turbidity conditions in estuaries, daily as well as seasonal, and to the small number of stations, the turbidity map (Figure A7) is indicative only. In particular, it says nothing about turbidity conditions in intertidal zones subject to tidal currents and wind-induced resuspension.

5.4 Primary Production

No production values for the Gulf were found in the literature. Hargraves and Viguez (1981) report biomass reaching 80 10^3 cells/ml for Cochlodinium catenatum, the main agent for red tides, which is apparently non toxic.

5.5 Metallic Pollutants

Dean et al. (1986) report heavy metal analyses for sediments and benthic fauna. Levels do not seem to exceed natural background, because of the absence of industrial pollution. The exception could be the vicinity of Puntarenas and the mouth of Tarcoles river in the outer gulf.

5.6 Water Quality of Tributaries

Rivers draining into the inner gulf have essentially agriculture and forested drainage basins. Pesticides and fertilizers are being used in these areas as per the list in Appendix 3, but they are not monitored in river water. Their input to the estuary therefore remains unknown. The water quality parameters analyzed show no sign of imbalance due to pollution, although inputs of nutrients are not known either.

6. MERGING OF GROUND AND SATELLITE INFORMATION

This operation would involve (a) calibration of TM imagery by means of ground data and (b) extrapolation of dry season conditions to wet season.

Part (a) would require field work coordinated with satellite overpasses. Since this is lacking in the present context, our interpretation of imagery remains qualitative. As for extrapolation, dry season conditions yield an optimistic picture of, at least, turbidity. The patterns observed on our images should shift towards an increase in turbidity all over the inner gulf, due to heavier freshwater discharge during the rainy season.

7. CONCLUSIONS

Bearing in mind the previous condiderations, the following conclusions can be reached:

(a) Remote sensing by means of Landsat TM "chromaticity" analysis makes possible the definition of water masses, characterized by mineral turbidity, planktonic blooms and bathymetry (bottom influence).

(b) In the absence of synoptic groundtruthing, quantitative calibration of the image in terms of turbidity was not possible with any degree of certainty.

(c) Whether the water mass patterns observed on the March and April images are permanent is not known.

(d) A trend map of maximum turbidity established on the basis of in situ measurements indicates total suspended solid concentration up to 100 mg 1-1 in the inner gulf, going down to 2-6 mg 1^{-1} off Puntarenas.

(e) Likewise, minimum salinity is less than 15 ppt N of Chira Island and regularly increases to about 30 ppt in the vicinity of Negritos Islands.

(f) For the time being, the innder gulf is not subject to heavy metal pollution, although the inputs of nutrients and pesticides remain unknown.

8. RECOMMENDATIONS

Water quality evaluation for aquaculture site selection in complex environments such as estuaries should include the following operations:

(a) Mapping of sediment typology in shallow water, using simple parameters such as percentage of sand, silt and clay fractions and organic matter content. More elaborate analyses of heavy metals and pesticides would also yield valuable information on water quality.

(b) Total suspended solids, Secchi depth and salinity measurements at the time of several satellite overpasses and also under the most extreme runoff or wind conditions.

(c) Development of a predictive remote sensing model of (at least) turbidity, in order to simulate a variety of conditions, thus leading to quantitative conclusions on the physico-chemical environment of aquaculture.

9. ACKNOWLEDGEMENTS

The author is grateful to Dr. W. Mooneyhan, Director and Mr. L. McGregor, both with GRID, for the facilities provided in the treatment of the imagery.

Appendix 1

Relationship between Total Suspended Solids, Secchi Depth and Beam Attenuation Coefficient

(Data of Klemas *et al.*, 1983)

1. Zs - c (Figure A8)

Only data for February 1981 were available, and Zs had to be estimated from Table 2 from Klemas *et al.* (Zs = sample depth). The relationship $c \simeq 4.5 / Zs$ was hand-fitted.

2. TSS - c (Figure A9)

February 1981 values were taken from Tables 2 and 11 from Klemas *et al.* (1983) at the surface and Secchi depths. Albeit scattered, the relationship is similar to that found in Chesapeake Bay by Bowker *et al.* (1975, Figure 9). It can be summarized by the statement: if TSS > 5 mg l^{-1} then $c \geq 5$ m^{-1}.

3. TSS - Zs (Figure A10)

Data from February and July 1981 are pooled on the figure, which shows the classical inverse logarithmic relationship (see for instance Lorenzen, 1980).

Appendix 2

Estimation of Depth of Light Penetration

1. White Light

Tyler (1968) proposes the following relation between Secchi depths Zs and attenuation coefficients:

$$Zs \simeq \frac{8.7}{c + K_v} \quad (1)$$

where Kv is the attenuation coefficient for illuminance and c the beam attenuation coefficient (Smith, Tyler and Goldman, 1973) in units of m^{-1}.

From data from Klemas *et al.* (1983) for the Gulf of Nicoya (Figure A8), we have derived the approximate relation:

$$c \simeq \frac{4.5}{Zs} \quad (2)$$

The penetration depth of satellite imagery has been defined by Gordon and McLuney (1975) as:

$$z_{90} \simeq \frac{1}{K_T} \quad (3)$$

Combining (1) and (2) gives:

$$K_v \simeq 0.93\ c \quad (4)$$

From Smith, Tyler and Goldman's (1973) data for lakes Crater and Tahoe, we can compute:

$K_T \cong 0.8$ kv (5)

and, from (4)

$K_T \cong 0.75$ c (6)

In the very turbid San Vicente reservoir (Tyler and Smith, 1970, one gets:

$K_T \cong 0.28$ Kv (7)

and, from (4)

$K_T \cong 0.26$ c (8)

The penetration depth can then be estimated from Secchi depths or c measurements as

$Z_{90} \cong 0.3$ Zs $\cong 1.3$ / c (9)

or

$Z_{90} \cong 0.85$ Zs $\cong 3.8$ / c (10)

Owing to the turbid character of Gulf of Nicoya water, relation (10) seems more appropriate.

2. Monochromatic Light

The penetration depth of satellite imagery has been computed by Gordon and McCluney (1975) as a function of wavelength and water type (according to Jerlov, 1968). This relationship is shown in Figure A11. Jerlov's typology is based on the spectral distribution of K_T, which is not available for Gulf of Nicoya waters.

However, we know that the relation (8) is valid for white light as well as at 530 nm. Hence K_T values for Nicoya can be backcalculated from (8) for extreme c values of 1 - 5 m^{-1} (see Appendix 1). They are plotted in Figure A12 at 530 nm, together with various water type spectra.

From this limited set of data, we tentatively conclude that Gulf of Nicoya waters are of coastal types 3 - 9. Penetration depth values for various TM bands can be estimated from Figure A11. They are summarized below.

Penetration Depth Z_{90} of TM Bands in Gulf of Nicoya (in meters)

TM Band Water type	1 Blue	2 Green	3 Red	4 NIR	White	ZS
Very turbid	1	1.5	1.2	1	.9	1
Turbid	4.5	4	2	<1	4.5	5
Type 1 (clear)	8	6	2.5	1.5	7	>5

Appendix 3

Summary on Agriculture Practice in Gulf of Nicoya Drainage Basin

(compiled by H. Nanne, Ministerio de Agricultura y Ganaderia de Costa Rica)

CULTIVOS MAS IMPORTANTES: (Important Crops)	Arroz (Rice) Pastos (Pasture) Sorgo (Sorghum) Algodón (Cotton)
PLAGUICIDAS: (Pesticides)	Propanil - herbicida 2-4 D " Metamidofos - Insecticida Metil paration " Mancozeb - fungicida Maneb " Tordon 101 (Picloran con 2-4D) Insecticidas Clorpirifos
FERTILIZANTES: (Fertilizers)	Nitrógeno - en fórmula 10-30-10 y 12-24-12 Fósforo " " " Potasio " " " Urea con 46% de nitrógeno

Figure A1. Landsat TM image. 15 April 1985. True colours. Western Gulf

Figure A2. Landsat TM image. 23 March 1985. True colours. Outer Gulf

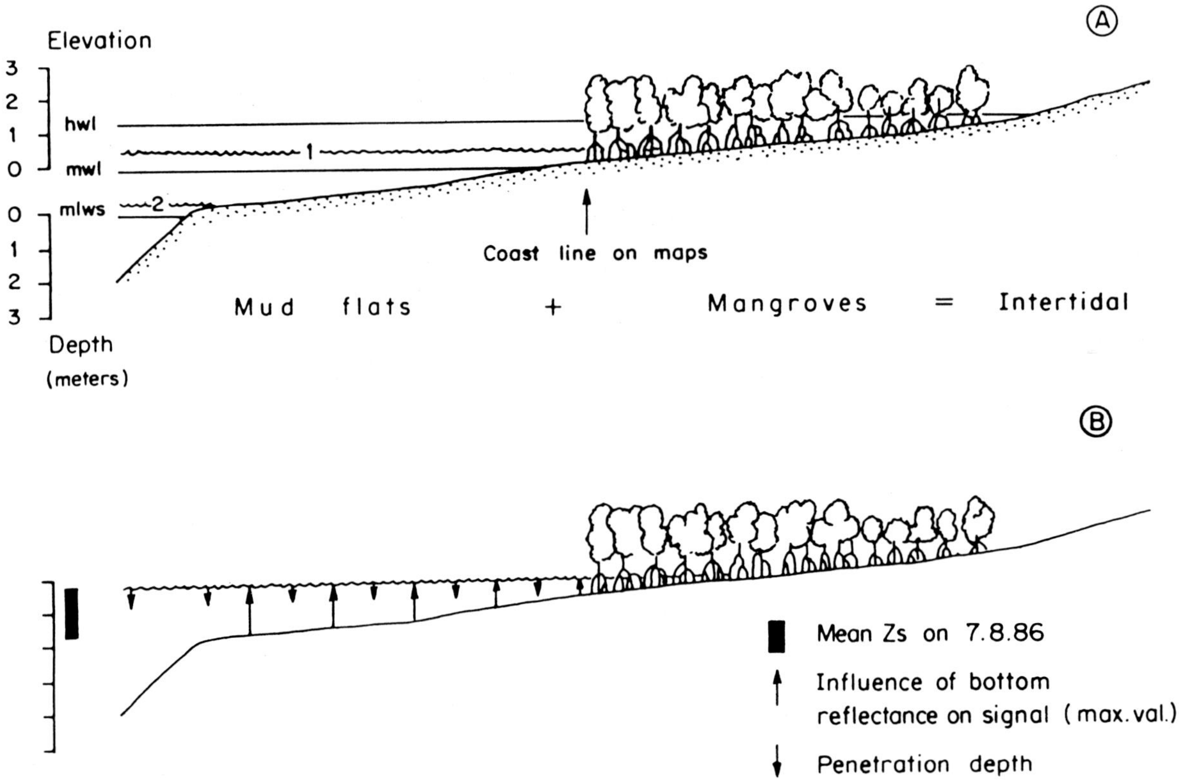

Figure A3. Coastal profiles in mangrove areas (vertical scale exaggerated)
A) Water level on 15.4.85 (1) and on 23.3.85 (2).
B) Penetration depth and bottom influence on signal based on Secchi readings (7.8.86) and Annex 2 formulas, for April image.

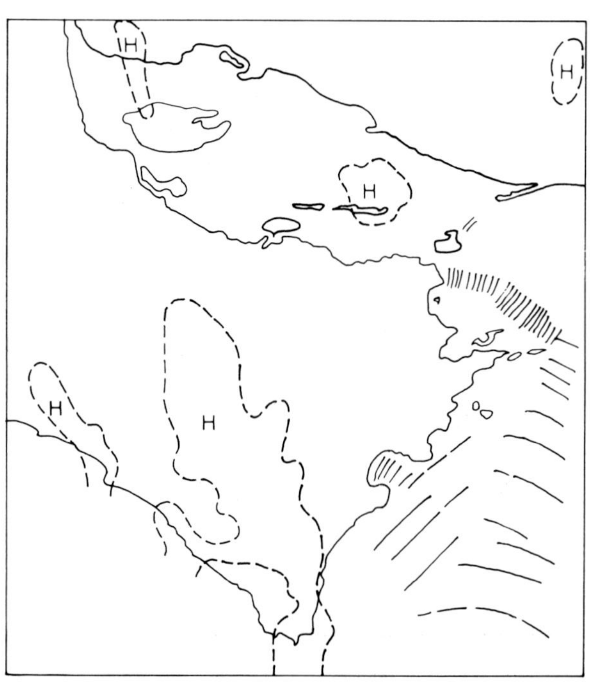

Figure A4a. Summary of water features. April image: haze and wave patterns

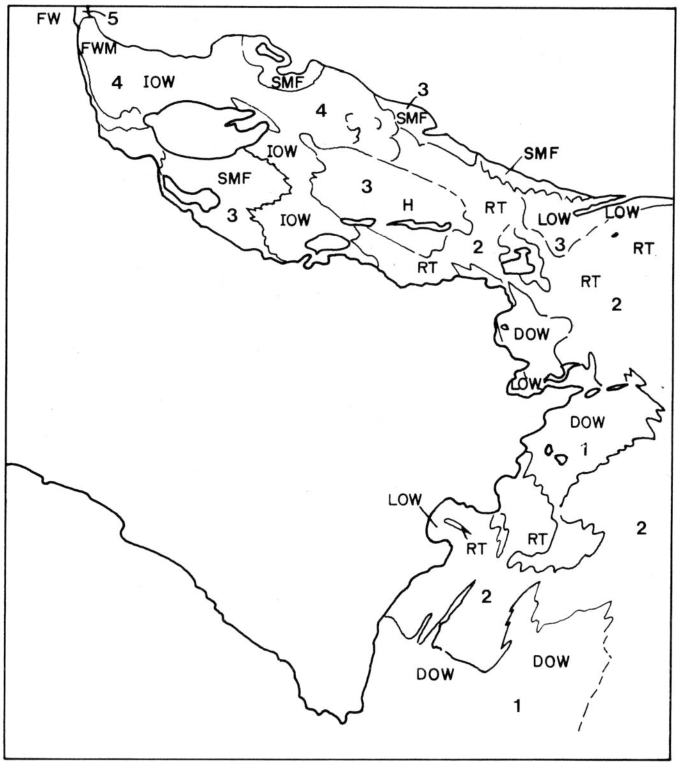

Figure A4b. Summary of water features. April image: water masses. 1: dark ocean water. 2: light ocean water with planktonic blooms. 3: turbid. 4: dark, greenish Upper Gulf water. 5: Rio Tempisque water

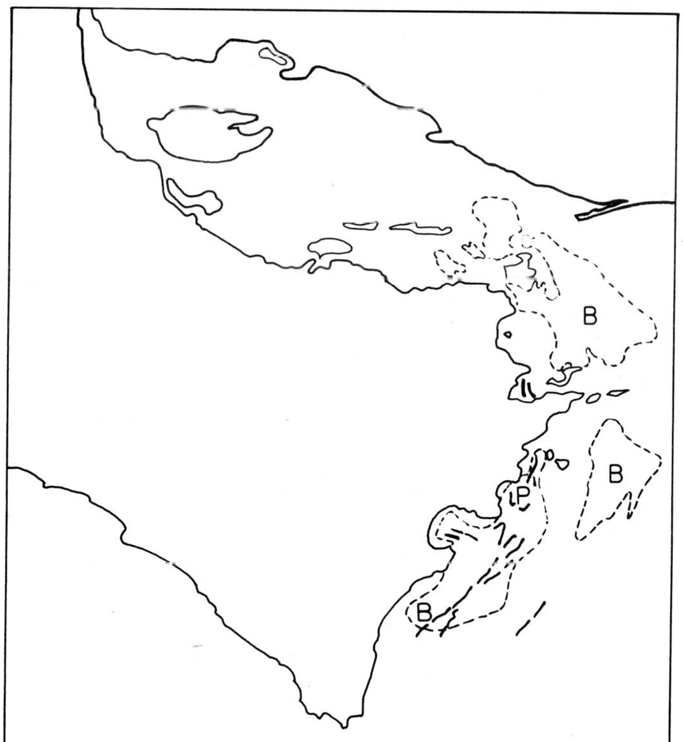

Figure A4c. Summary of water features. April image: planktonic blooms. B: brown tides. P: pink tides. Dark slicks also shown as thick black lines

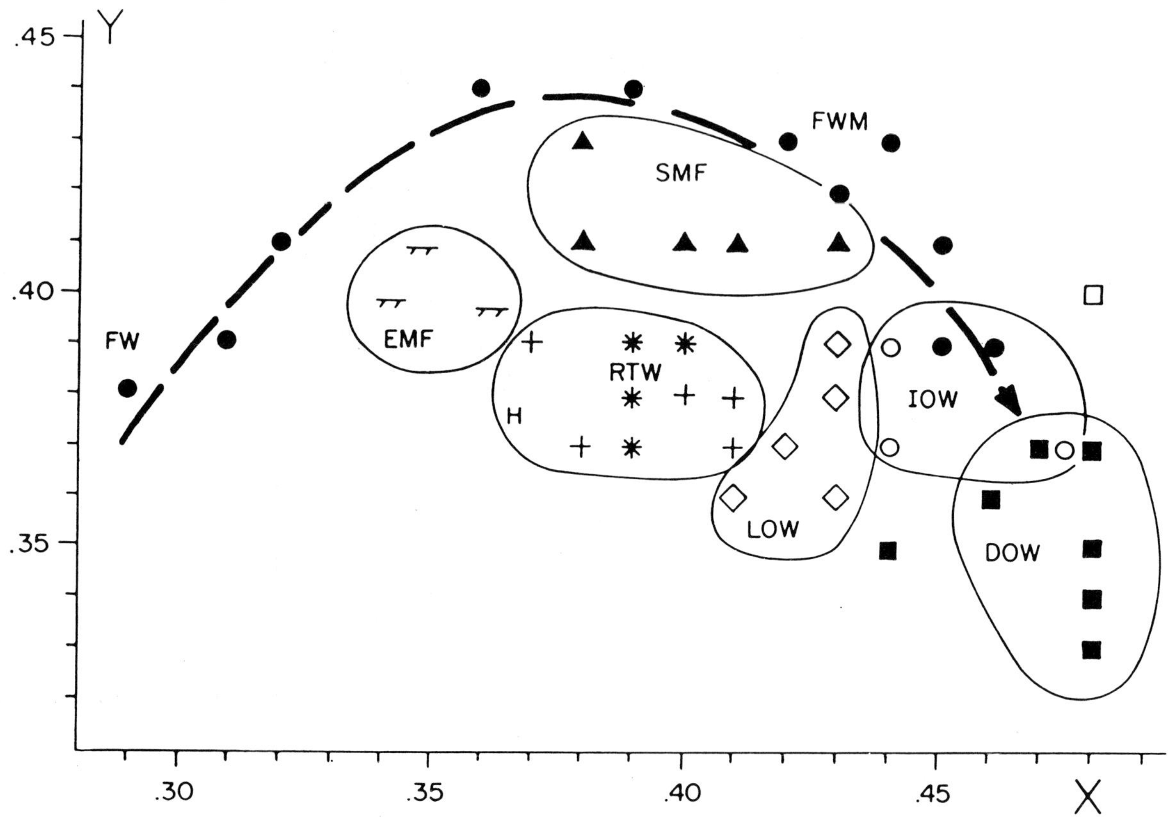

Figure A5. "Chromaticity" diagram for selected points. April

Rio Tempisque. FW: purely fluviatile. FWM: mixed

Upper Gulf (Chira). IOW: Isolated Ocean Water

Submerged mud flats (SMF)

Emerged mud flats (EMF)

Dark Ocean Water (DOW)

Light Ocean Water (LOW)

Red/brown tides (RTW)

Haze (H)

— — → Inferred turbidity gradient

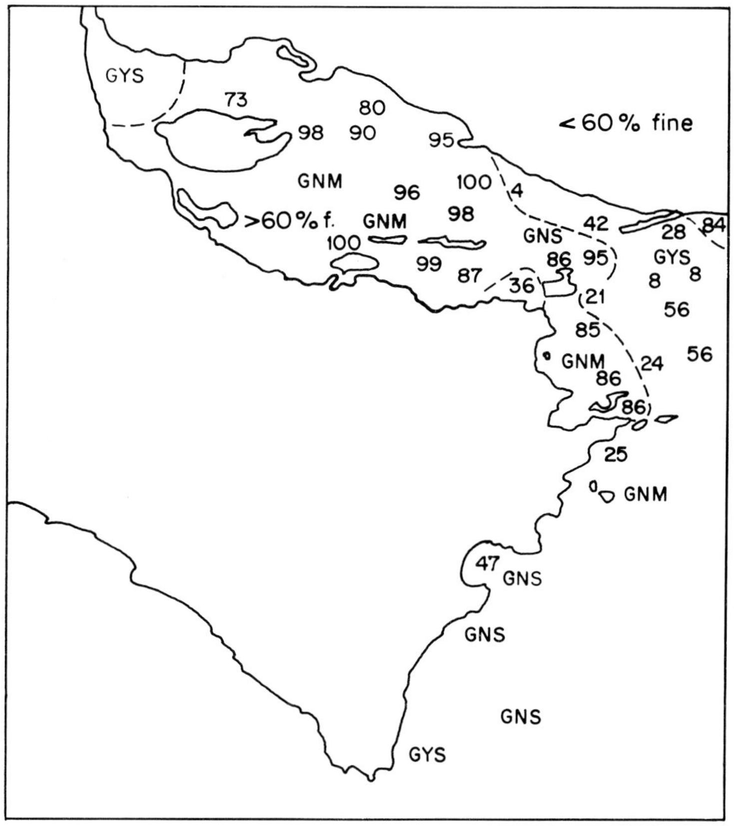

Figure A6a. Bottom-sediment types: percentage of size fraction <62 um and sediment type. GNM - green mud; GNS green sand; GYS - grey sand

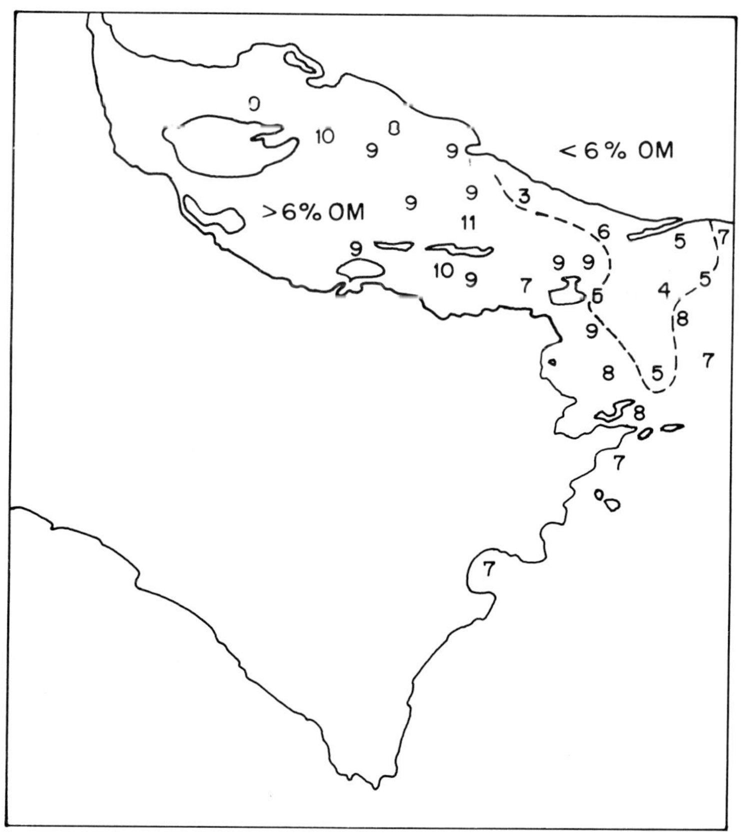

Figure A6b. Bottom-sediment types: percentage of organic matter (after Universidad de Costa Rica, 1985)

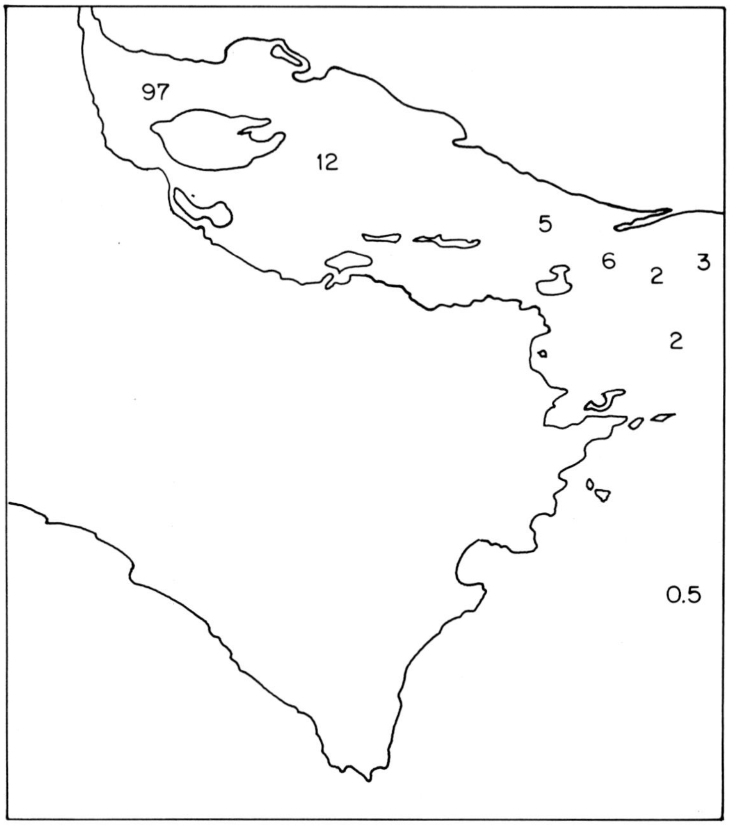

Figure A7. Maximum-turbidity map (after Klemas et al., 1983). Values in mg l^{-1} TSS

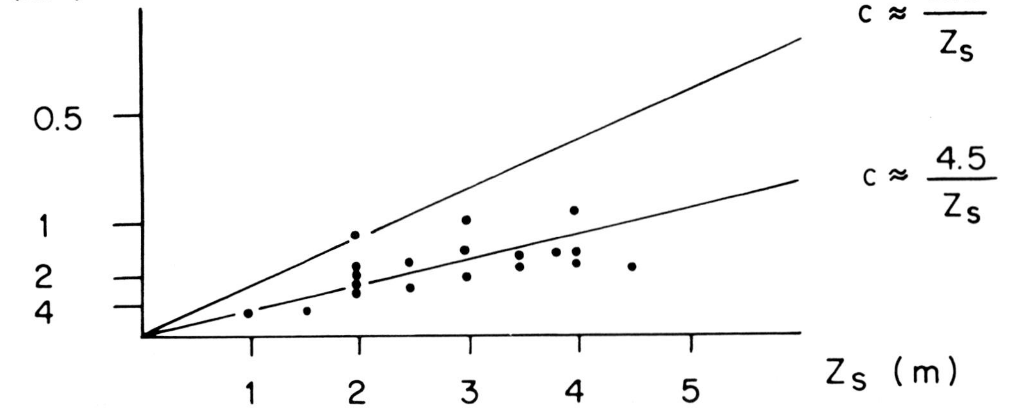

Figure A8. Empiricial relationship between Secchi Depth (Zs) and Beam Attenuation Coefficient (c) for Gulf of Nicoya February 1981 data (after Klemas et al., 1983)

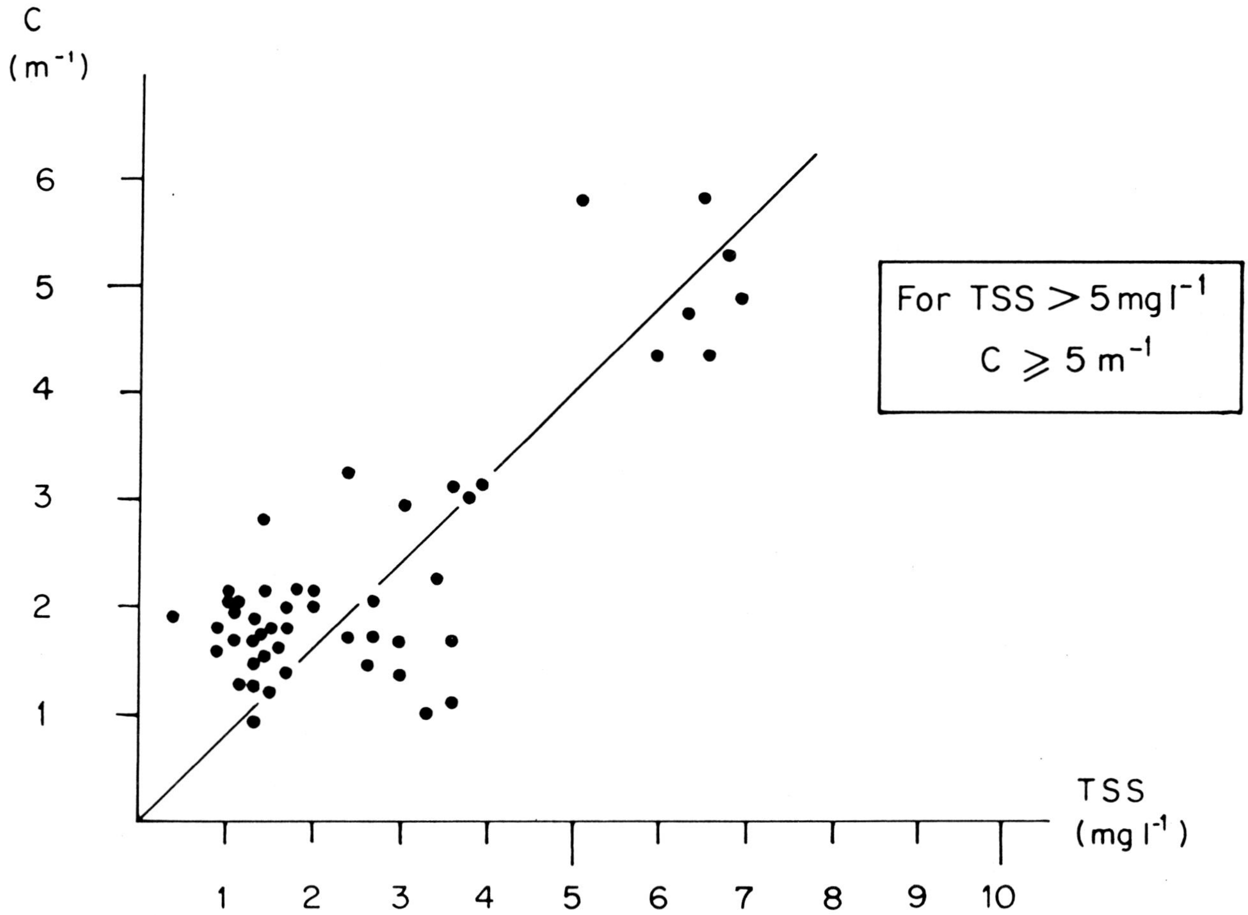

Figure A9. Empirical relationship between Total Suspended Solid concentration (TSS) and Beam Attenuation Coefficient (c) for Gulf of Nicoya February 1981 data (after Klemas et al., 1983)

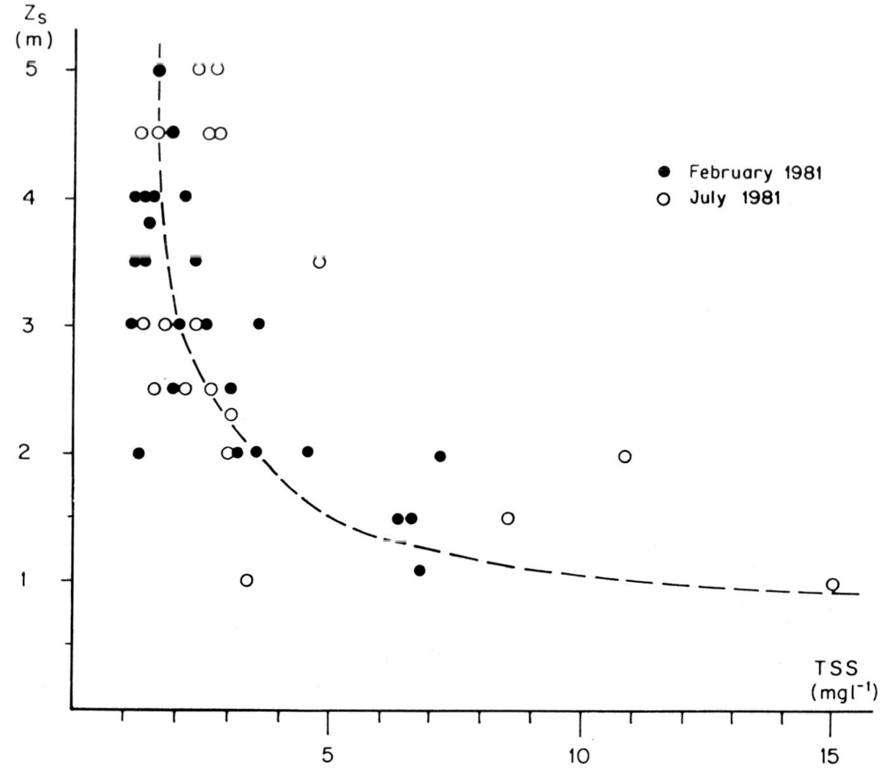

Figure A10. Empiricial relationship between Total Suspended Solid concentration (TSS) and Secchi Depth (Zs) for Gulf of Nicoya February and July 1981 data (after Klemas et al., 1983)

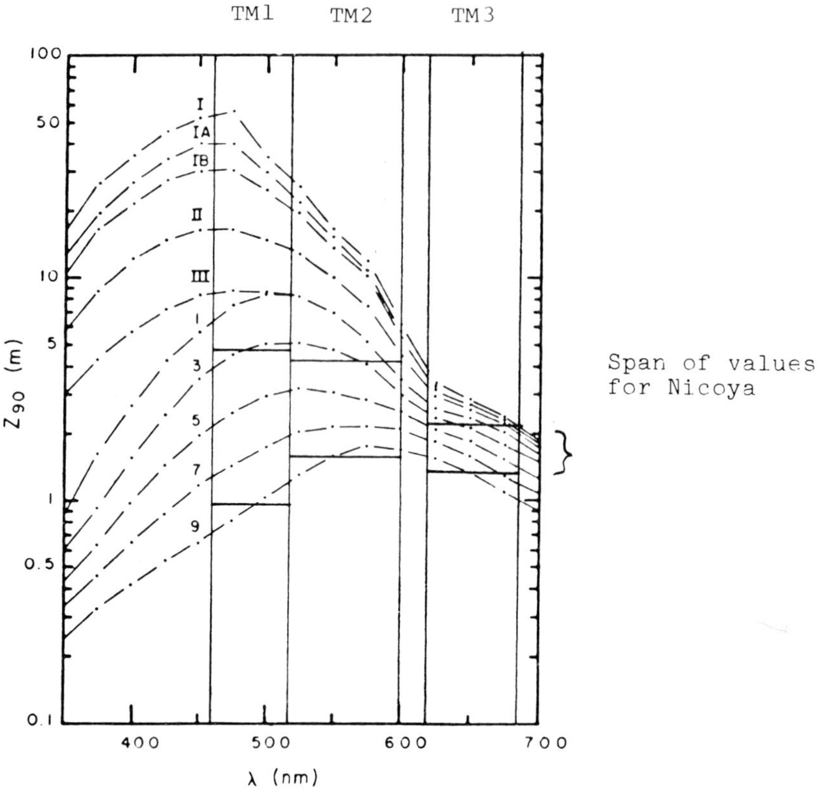

Figure A11. Variation of Z_{90} with wavelength and TM band for various water types (modified from Gordon and McCluney, 1975)

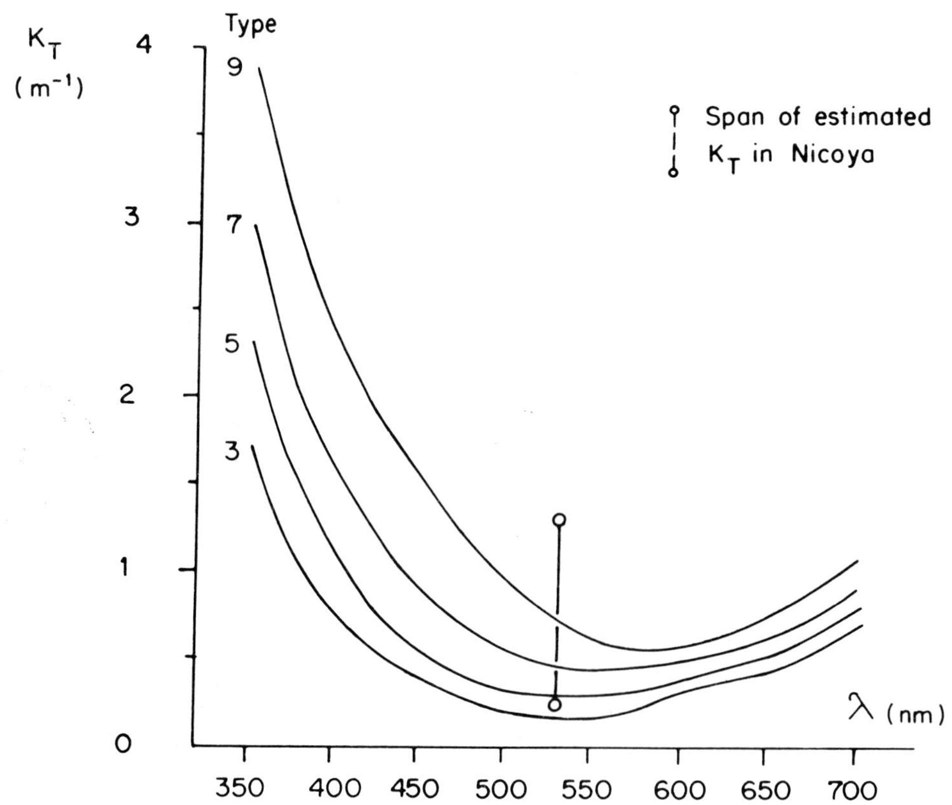

Figure A12. Spectral variation of K_T for Type 3 - 9 waters and span of estimates K_T in Gulf of Nicoya (after Jerlov, 1968, p.135)

Annex 2

Inner Gulf Solar Salt Ponds: UTM Coordinates, Latitude and Longitudes, and Surface Areas

	EASTING	NORTHING	LONGITUDE	LATITUDE	AREA
1	693150	1132800	85 14 11	10 14 36	3.15
2	693390	1132440	85 14 3	10 14 25	1.17
3	693750	1132230	85 13 52	10 14 18	4.86
4	693990	1131150	85 13 44	10 13 42	1.44
5	694800	1129620	85 13 17	10 12 53	2.79
6	695250	1129560	85 13 3	10 12 51	2.34
7	694680	1129410	85 13 22	10 12 46	5.94
8	694380	1129110	85 13 31	10 12 36	5.67
9	696960	1128060	85 12 7	10 12 1	4.68
10	700260	1126440	85 10 19	10 11 8	1.53
11	708300	1126290	85 5 55	10 11 2	11.07
12	709470	1126050	85 5 16	10 10 54	24.39
13	710400	1126230	85 4 46	10 10 59	1.17
14	708810	1126110	85 5 38	10 10 56	0.99
15	696900	1126080	85 12 9	10 10 57	39.42
16	710070	1126020	85 4 57	10 10 53	4.95
17	697890	1126110	85 11 37	10 10 58	5.58
18	702000	1125840	85 9 22	10 10 48	18.63
19	708150	1125900	85 6 0	10 10 49	1.89
20	703530	1125870	85 8 31	10 10 49	1.53
21	711330	1125870	85 4 15	10 10 47	1.08
22	701430	1125780	85 9 40	10 10 46	1.53
23	704160	1125720	85 8 11	10 10 44	10.26
24	705060	1125570	85 7 41	10 10 39	29.16
25	707430	1125690	85 6 23	10 10 42	8.46
26	711630	1125540	85 4 5	10 10 37	3.33
27	698040	1125450	85 11 32	10 10 36	2.16
28	711900	1125420	85 3 57	10 10 33	2.25
29	691560	1125090	85 15 5	10 10 26	1.08
30	691560	1124010	85 15 5	10 9 51	2.79
31	702960	1124130	85 8 50	10 9 52	2.43
32	712380	1123890	85 3 41	10 9 43	4.95
33	691710	1123800	85 15 0	10 9 44	1.62
34	691560	1123200	85 15 5	10 9 24	10.98
35	691110	1122570	85 15 20	10 9 4	20.79
36	712470	1122870	85 3 38	10 9 10	8.28
37	690330	1122510	85 15 46	10 9 2	3.60
38	713400	1122300	85 3 8	10 8 51	7.38
39	717300	1120260	85 1 0	10 7 44	5.58
40	697980	1118760	85 11 35	10 6 59	4.77
41	689460	1118700	85 16 15	10 6 58	2.43
42	690180	1118160	85 15 51	10 6 40	0.90
43	704880	1118070	85 7 49	10 6 35	9.45
44	690120	1118010	85 15 53	10 6 36	1.62
45	690540	1117470	85 15 40	10 6 18	1.98
46	690840	1117140	85 15 30	10 6 7	11.16
47	692820	1116960	85 14 25	10 6 1	3.24
48	723060	1116150	84 57 52	10 5 29	6.75
49	723480	1115760	84 57 38	10 5 16	3.96
50	724350	1115340	84 57 10	10 5 2	0.99
51	724740	1115310	84 56 57	10 5 1	2.07
52	725250	1115100	84 56 40	10 4 54	6.93
53	725250	1114650	84 56 40	10 4 40	3.60
54	695400	1114500	85 13 1	10 4 40	10.80
55	725190	1114110	84 56 42	10 4 22	8.82
56	728640	1113600	84 54 49	10 4 5	38.61
57	724860	1113600	84 56 53	10 4 5	2.43
58	725310	1113420	84 56 39	10 3 59	5.49
59	724380	1113300	84 57 9	10 3 56	1.53
60	727050	1111560	84 55 42	10 2 59	9.36
61	696750	1111440	85 12 17	10 3 1	7.11
62	700080	1110750	85 10 28	10 2 37	3.60
63	729300	1110300	84 54 28	10 2 17	16.92
64	697800	1108290	85 11 43	10 1 18	7.47
65	699060	1107900	85 11 2	10 1 5	11.70
66	699720	1107390	85 10 40	10 0 48	12.42
67	699330	1107300	85 10 53	10 0 45	4.50
68	701790	1106970	85 9 32	10 0 34	22.86
69	702690	1106190	85 9 3	10 0 9	4.59
70	704010	1104780	85 8 20	9 59 22	73.62
71	704940	1103370	85 7 49	9 58 36	3.15
72	712980	1102590	85 3 26	9 58 9	3.69
73	707610	1102560	85 6 22	9 58 10	26.82
74	708300	1102200	85 5 59	9 57 58	10.44
75	715740	1101540	85 1 55	9 57 35	7.65
76	716700	1100490	85 1 24	9 57 0	9.72
77	726090	1093800	84 56 17	9 53 21	1.80
78	726960	1088940	84 55 49	9 50 43	1.80
79	726750	1087680	84 55 57	9 50 2	13.86
80	729150	1085100	84 54 38	9 48 37	3.24

Tip. INTERSTAMPA r.l. - Via Barbana, 33 - Tel. 06/540.33.49